Cold War Civil Defence in Western Europe

Marie Cronqvist · Rosanna Farbøl ·
Casper Sylvest
Editors

Cold War Civil Defence in Western Europe

Sociotechnical Imaginaries of Survival and Preparedness

palgrave
macmillan

Editors
Marie Cronqvist
Department of Communication and Media
Lund University
Lund, Sweden

Rosanna Farbøl
Department of History
University of Southern Denmark
Odense, Denmark

Casper Sylvest
Department of History
University of Southern Denmark
Odense, Denmark

ISBN 978-3-030-84283-3   ISBN 978-3-030-84281-9 (eBook)
https://doi.org/10.1007/978-3-030-84281-9

This Palgrave Macmillan imprint is published by the registered company Springer Nature Switzerland AG
The registered company address is: Gewerbestrasse 11, 6330 Cham, Switzerland

Preface

The idea for this book emerged at a workshop in the Network for Civil Defense History held at Lund University in the autumn of 2018. In the delightful surroundings of the Old Bishop's House, our conversations brimmed with observations about the similarities and differences in national civil defence efforts during the Cold War and theoretical discussions about how to approach and study such histories. Notwithstanding the interdisciplinary composition of the group, there was agreement that the field of civil defence history would benefit from a common theoretical vocabulary and a sustained effort in the direction of transnational and comparative history. It was against this background that we decided to collectively explore the merits and potential of the concept of sociotechnical imaginaries in a branch of history so deeply affected by the development of science and technology. When we met again, this time in wonderful Copenhagen at the height of summer in August 2019, most of the chapters began to take shape. We thank Riksbankens Jubileumsfond and the Danish Research Council (grant no. 8018-00047B, Danish Civil Defence during the Cold War) for their support in making these workshops possible.

Shortly after lockdowns were introduced in many countries during the spring of 2020, we moved our collaboration online, submitting and commenting on drafts through a web-based collaborative platform. The pandemic and its waves have had deep effects on societies, communities, families and individuals across the world. For teachers and scholars of history, neat plans and schedules were disrupted due to irregular access to archives and new forms of teaching that encroached on research time. What stands out, however, from our perspective as editors, is just how rewarding and gratifying scholarly collaboration has proved to be during this time. We cannot thank the contributors to this volume enough for the spirit and energy with which they have taken on this task under novel circumstances.

It has been our priority from the outset to make this volume as widely accessible as possible, and we are deeply grateful to the Book Fund at the Lund University Library for covering the Open Access fee, and thereby making the entire volume available to access and download via the Palgrave Macmillan website. We would also like to thank Den Hielmstierne-Rosencroneske Stiftelse for its financial support, which has proved extremely valuable in preparing the manuscript for publication. In that context, we would also like to express our gratitude to Catherine Schwerin, who has improved the manuscript with great care and a sharp eye for every detail. Finally, we thank Molly Beck, Lucy Kidwell and Joe Johnson at Palgrave Macmillan for their interest in the project and their support and professional assistance in all of its phases.

Lund, Sweden
Aarhus, Denmark
Odense, Denmark
April 2021

Marie Cronqvist
Rosanna Farbøl
Casper Sylvest

Contents

Notes on Contributors

Peter Bennesved is a Historian of science and ideas at the University of Umeå, Sweden. His research concerns Swedish civil defence during the twentieth century, with a particular focus on its material aspects and technologies. In November 2020, Bennesved defended his thesis *Sheltered society civilian air raid shelters in Sweden—from idea to materiality, 1918–1940 and beyond* (2020). Bennesved has also conducted research on Cold War civil defence media and propaganda. Currently, he is working on the reception of internet, surveillance and cybersecurity technologies in co-operation with the Humanities Laboratory (HUMLAB) and the *iAccept* project at Umeå University.

Iben Bjørnsson is a Researcher and Curator at Stevnsfort Cold War Museum in Denmark specialising in contemporary history. She earned her Ph.D. in History from the University of Copenhagen and has done extensive work on cultural, ideological, military and political aspects of the Cold War, security and international relations, political parties and organisations, the Arctic and the Nordic countries. She is the author of a biography on Copenhagen's former lord mayor Urban Hansen and a book on the anti-communist activities of the Danish Social Democratic party.

Marie Cronqvist is Associate Professor in Journalism and Media History at the Department of Communication and Media at Lund University, Sweden. Her research is focused on Cold War media, civil defence culture and transnational broadcasting. Her most recent publications include 'Foreign correspondents in the Cold War: The politics and practices of East German television journalists in the West' (co-author Sune Bechmann Pedersen) in *Media History* (2020) and 'From socialist hero to capitalist icon: The cultural transfer of the East German children's programme Unser Sandmännchen to Sweden in the early 1970s' in *Historical Journal of Film, Radio and Television* (2021).

Rosanna Farbøl is a Postdoctoral Researcher at the University of Southern Denmark. Her research is focused on Cold War civil defence, and she has specialised in the connection between materiality, imaginaries and culture, and comparative studies of urban civil defence strategies in Denmark and Europe. Her latest publications include 'Warfare or welfare? Civil Defence and emergency planning in Danish urban welfare architecture' in *Cold War cities: Politics, culture and atomic urbanism*, eds. Jonathan Hogg, Martin Dodge and Richard Brooks (London: Routledge, 2020) and 'Urban civil defence: Imagining, constructing and performing nuclear war in Aarhus', *Urban History* (2020).

Matthew Grant is a Reader in History at the University of Essex, United Kingdom. He has written extensively on British civil defence, including *After the bomb: Civil defence and nuclear war in Britain, 1945–68* (Palgrave, 2010) and several articles and essays. With Benjamin Zieman, he co-edited *Understanding the imaginary war: Culture, thought, and nuclear conflict* (Manchester University Press, 2016). He is currently writing a history of *Britain's Cold War home front: Citizens and the state, 1945–53.*

Jonathan Hogg is Senior Lecturer in Twentieth-Century History at the University of Liverpool. He researches the cultural and social history of British nuclear mobilisation. He is the Author of *British nuclear culture: Official and unofficial narratives in the long twentieth century*(Bloomsbury Academic, 2016) and the Editor or co-editor of *Using primary sources* (LUP, 2017), *Cold War cities* (Routledge, 2020) and *British nuclear mobilisation since 1945* (Routledge, 2021). He is currently working on histories of UK uranium mining and on an extended project on the nuclear 1980s.

Sibylle Marti is a Postdoctoral Researcher at the University of Bern, Switzerland. Her research is located at the interface of cultural history, history of knowledge and STS. She has published widely on the history of the Cold War and co-edited volumes on the imaginary of the Cold War in Europe as well as on Switzerland's armament culture during the Cold War. Her most recent book entitled *Radiation in the Cold War* (2020, *Strahlen im Kalten Krieg*) investigates the history of radiation research and radiation protection in Cold War Switzerland.

Jochen Molitor is a Former Researcher at the University of Cologne, Germany. His work is focused on contemporary West German history, with a special emphasis on emergency management and preparedness in relation to nuclear weapons and a variety of other risks. He is the Author of *Surviving the bomb* (2011, *Mit der Bombe überleben*), a study of civil defence literature and communication during the early 1960s, and several articles on the wider history of civil defence in the Federal Republic of Germany. His dissertation on the history of West German disaster medicine, "Disaster as profession" (Katastrophe als Beruf), was published in September 2021.

Casper Sylvest is Associate Professor in History at the University of Southern Denmark. His research is focused on the cultural and intellectual reception and processing of nuclear weapons technologies during the Cold War, especially in Denmark, Britain and the USA. He is currently directing a research project on Danish civil defence (DACIDE). His recent publications include 'Pre-enacting the next war: The visual culture of Danish civil defence in the early nuclear age' in *Cold War History* (2021) and 'Nuclear weapons, extinction, and the Anthropocene: Reappraising Jonathan Schell' (with Rens van Munster) in *Review of International Studies* (2021).

Dick van Lente is a Retired Lecturer in cultural history at Erasmus University in Rotterdam. He participated as an editor and contributor in two-book series on the history of technology in the Netherlands in the nineteenth and the twentieth centuries. The final volume of the latter series appeared as *Technology and the making of the Netherlands* (MIT Press, 2009). He has edited *The nuclear age in popular media* (Palgrave Macmillan, 2012). He is currently finishing an edited volume about the cultural reception of computing in 12 countries around the world. He serves the journal *Technology and Culture* as review editor.

Abbreviations

BB	Bescherming Bevolking (Protection of the Population, The Netherlands)
BLSV	Bundesluftschutzverband (West Germany)
BzB	Bundesamt für zivilen Bevölkerungsschutz (West Germany)
CDC	The NATO Civil Defence Committee
DCDD	Danish Civil Defence Directorate (Denmark)
DCDL	Danish Civil Defence League (Denmark)
FCDA	Federal Civil Defense Administration (USA)
FVVH	Nederlandse Federatie voor Vrouwelijke Vrijwillige Hulpverlening (Dutch Federation of Women's Voluntary Organisations, The Netherlands)
IPPNW	International Physicians for the Prevention of Nuclear War
PSR	Physicians for Social Responsibility (Switzerland)
SACEUR	Supreme Allied Commander Europe (NATO)
THW	Bundesanstalt Technisches Hilfswerk (West Germany)
UVV	Unie van Vrouwelijke Vrijwilligers (Union of Women Volunteers, The Netherlands)

List of Figures

CHAPTER 1

Introduction: New Paths in Civil Defence History

Marie Cronqvist, Rosanna Farbøl, and Casper Sylvest

For a number of years after the end of the Cold War and the proclaimed ascendancy of Western liberal democracy, nuclear civil defence and emergency preparedness were considered by many to be a relic of history, a curiosity or vaguely familiar target of ridicule. In the 2020s, this is no longer the case. As expressions of fear and a desire for survival and resilience should the worst happen, institutions and practices of civil defence have both predated and outlived the Cold War. In the age of terrorism, global pandemics, rising political tensions between nuclear armed states, and a deepening climate crisis, emergency preparedness

M. Cronqvist (✉)
Department of Communication and Media, Lund University, Lund, Sweden
e-mail: marie.cronqvist@kom.lu.se

R. Farbøl · C. Sylvest
Department of History, University of Southern Denmark, Odense, Denmark
e-mail: rosfa@sdu.dk

C. Sylvest
e-mail: csy@sdu.dk

M. Cronqvist et al. (eds.), *Cold War Civil Defence in Western Europe*,
https://doi.org/10.1007/978-3-030-84281-9_1

is once again making headway on the political agenda in many countries. Continuities between Cold War civil defence and new logics of (in)security (Masco 2013; Deville et al. 2014; cf. Collier and Lakoff 2008) have become apparent, and authorities, politicians and defence intellectuals seem to be increasingly interested in updating material infrastructures and revisiting core ideas of civil defence.

In a striking example of the Cold War echoing in our own time, in May 2018 the Swedish Government distributed a leaflet entitled *If crisis or war comes* to every household in Sweden. It highlighted the importance of emergency preparedness, asking citizens to prepare and stockpile food and necessities and instructing the reader to 'educate yourself!', while providing the location of nearby shelters (MSB 2018). By using an expression already established in the 1940s and repeated constantly throughout the postwar era—'If the war comes'—the Swedish government today is forging a peculiar connection between the Cold War nuclear mindset and twenty-first-century narratives of fear and crises. In Europe, new initiatives are also being introduced by the European Commission in its multibillion-euro research programme *Horizon Europe*, a part of which prioritises research into 'Civil Security for Society'. Apart from improving situational awareness, data-driven decision-making and the understanding of risk exposures among authorities, the Commission seeks ways to cultivate a '"culture of disaster preparedness" for citizens, communities, public administrations, business companies, practitioners…'.[1]The COVID-19 pandemic, which began in 2020 and has wreaked havoc across the globe, has further fuelled demands for emergency preparedness in societies that aim to be resilient in the face of a wide variety of risks. In particular, the dilemma of how to communicate appropriately with the public in times of crisis has multiple parallels in the history of civil defence. Old technologies have also been revived. The pervasive climate crisis has, for example, brought new attention to the use of shelters as vital protection from cyclones, hurricanes and wildfires, concluding a shift that began during the late Cold War away from threats coming from a 'stable enemy' and towards an all-hazards approach to threats posed by 'nonspecific adversaries' (Lakoff 2007).

At the same time, scholars are taking a new interest in risk, resilience and emergency management. A recently funded American project, 'Reinventing civil defence', seeks to develop 'effective non-partisan nuclear risk communication targeted at Millennials' in an effort to inspire change in nuclear weapons politics (Karl and Wellerstein 2017; Karl and Lytle

2019). Meanwhile, within the developing and highly publicised field of Global Existential Risks, the value of civil defence thinking and material structures like refuges and shelters is being revisited with a view to assessing their importance for ensuring the survival of humanity in the face of new all-encompassing threats like super volcanoes or mismanaged geoengineering projects.[2]

As we enter an era in which civil defence and crisis preparedness measures are on the rise, it is key to remember past lessons. One motivation for this book lies in its aligning of historical research and contemporary perspectives. Planning, executing and experiencing institutionalised forms of emergency preparedness did not historically—nor does it today—take place in isolation from societal, cultural and political agendas. The chapters in this book also testify to the intricate and complex entwinement of civil defence with regional, national and global contexts. Historical knowledge is relevant to understanding the logic, results and consequences of contemporary attempts to secure the lives of citizens. Thus, we must return to civil defence and study its modes of operation, its material expressions, its underlying policies and mindsets, and not least its imaginary, performative qualities. How were civil defence objectives conceived, how were they implemented and with what consequences for societies and everyday existence? What forms of power were embedded in these systems and practices and how were they exercised?

Historiography, Trends and Aims

Cold War civil defence already attracted the attention of political and cultural historians in the 1980s (Boyer 1985; Vale 1987; May 1988), but the historiography developed significantly after the end of the Cold War. For a long time, the research field generally focused on conditions in the USA, which coalesced with a more general interest on the part of American contemporary historians in looking back on Cold War culture or 'atomic/nuclear culture' from the vantage point of hindsight.[3] Studies by Guy Oakes (1994) and Tracy C. Davis (2007) have focused on the entanglement of civil defence, administration and bureaucracy, Kenneth D. Rose (2001) and David Monteyne (2011) have researched the American shelter system, and Laura McEnaney (2000) has demonstrated how the home became a Cold War front line as civil defence contributed to a militarisation of domestic space. Recently, the first truly comparative study of the nature of civil defence in the USA and the Soviet Union has appeared (Geist 2019).

When European researchers developed a sustained interest in the history of civil defence, gradually from the early 2000s, one side-effect was an expanded chronological scope. From the clear focus on the 1950s that characterised the scholarship on American Cold War culture, later decades, not least the 1960s and 1980s, began to receive attention. Though the perspectives changed, the themes largely followed those in the established American research field: the organisational and political rationale behind civil defence (Hennessy 2002; Grant 2010), civil defence literature and propaganda (Molitor 2011; Cronqvist 2012; Stafford 2011), the role of cultural norms and emotions in mobilising the citizens around national values (Cronqvist 2015; Biess 2009; Hogg 2016), bunkers and shelter construction (Bennett 2011; Berger Ziauddin 2017b; Bennesved 2020) and the role played by scientific uncertainty about the properties and of radiation in public appraisals of civil defence (Bennett 2018; Sylvest 2020a).

Taken together, the field of civil defence history today displays a rich and productive mix of political and cultural historical perspectives following the so-called cultural turn in the 1990s. While much has been achieved in terms of charting the political significance and cultural dimensions of civil defence, especially within national frameworks,[4] important dimensions are still missing. With this edited volume, we aim to build on previous research on civil defence by actively seeking to integrate three central dimensions.

First, the role of technology, science and materiality in preparing for and rehearsing the worst warrant more attention. For decades historians have taken an interest in science and technology, as the voluminous subfields of the history of science and the history of technology testify, and the study of material culture is now also well integrated into historical scholarship. Yet, these aspects of history and historical experience have not so far been systematically explored in civil defence history. Several scholars have taken important steps in this direction,[5] but the field of civil defence history would benefit from a more sustained and theoretically informed focus on these themes.

Second, transnational and comparative perspectives in civil defence history are long overdue. As a subfield of history—a discipline with strong ties to the nation—dealing with a core task of the state, defence, it is hardly surprising that it has remained overwhelmingly national in orientation. Still, it is striking how limited our knowledge is about variations both within and between countries when it comes to how civil defence developed as a response to the threat of nuclear war and the effects it had on societies, politics, cultures and citizens. Moreover, existing scholarship has not paid enough attention to the actors and arenas that were central for the transnational circulation of ideas about civil defence and emergency planning. They include international organisations like NATO but also more informal ties between civil defence officials or professionals in allied and neutral countries.

Third and finally, we seek to bring more attention to the divergent and various everyday experiences of civil defence in different contexts. These contexts are not only national, but sometimes also—and perhaps more importantly—local or regional.[6] Experiences of civil defence also include imaginaries and narratives about Cold War civil defence that inform collective memory, as well as affective patterns or structures that actively contribute to contemporary views of civil defence and the forms of emergency management now enjoying political attention. What Cold War imaginaries linger on in our everyday lives today?

It is in dialogue with the abovementioned historiography on civil defence that this volume seeks to formulate new ambitions for the field and bring together scholarship on Europe during the Cold War. We aim to advance the field by collecting histories of civil defence within an analytical framework that (1) enhances our understanding of similarities and differences in national civil defence efforts as well as in local and transnational dynamics and (2) places special emphasis on the interplay between technology, materiality, culture and politics in civil defence. Individually and collectively, the chapters in this book pursue these aims, and in doing so they operate on three levels of analysis that are highlighted in the organisation of the volume.

The geographical scope of the book is centred on North-Western Europe. We assemble and seek to integrate historical scholarship on civil defence in six European countries—Denmark, Sweden, the Netherlands, West Germany, Switzerland and the UK—and one international organisation, NATO. While the majority of these countries were united by NATO membership and a transatlantic bond, several chapters focus on the role of neutral countries. Sweden and Switzerland are important to include because of their extensive efforts and prominent status in civil defence. After all, these countries allocated the largest funding per capita to civil defence in the postwar era (Vale 1987). It is clear that the countries that invested the most in civil defence in the Cold War were indeed those that culturally, materially and financially were not too bruised by the Second World War (Cronqvist 2012).

Chronologically, the focus of the book is the Cold War. However, in contrast to much previous research on Cold War civil defence, our time frame extends well beyond the 1950s and early 1960s in order to grasp some of the interesting trajectories and discrepancies not only between different countries, but also between different periods composing the postwar era. The Cold War is thus more an empirical time frame than a concept with inherent and distinct qualities across more than four decades. As will be apparent in the chapters that follow, other chronologies appear that are equally key to understanding civil defence in its historically and socially situated contexts. Some histories will need to be anchored in the Second World War or the interwar period, while others will resonate with present-day visions of civil defence.

Sociotechnical Imaginaries: Introducing and Exploring a Concept

The ambitions of this volume place several demands on our analytical approach. Apart from being attentive to the role of science and technology in social life and being open to transnational dynamics, it also aims to capture some of the most fascinating aspects of civil defence as policy, materiality, practice and vision while being able to accommodate a wide range of themes that coalesced in the institutions, practices and norms of civil defence. We argue that the concept of *sociotechnical imaginaries* developed by Sheila Jasanoff and Sang-Hyun Kim (2015) provides a productive starting point for exploring new paths in civil defence historiography. In its latest incarnation, Jasanoff has defined sociotechnical imaginaries as

> collectively held, institutionally stabilized, and publicly performed visions of desirable futures, animated by shared understandings of forms of social life and social order attainable through, and supportive of, advances in science and technology. (Jasanoff 2015a: 4)

The underlying logic of sociotechnical imaginaries is based on the idiom of co-production, a way of seeing ‘knowledge and its material embodiments' as 'at once products of social work and constitutive of forms of social work' (Jasanoff 2004: 2). The concept has been developed with the explicit ambition of uniting two views of modernity: a focus on utopias and dystopias prevalent in much political theory (and cultural studies) and a sophisticated understanding of the role(s) of science and technology in social life developed, above all, within Science and Technology Studies (STS).

Apart from a strong attraction to social theory, STS is characterised by a sustained interest in specifying how science and technology matters.[7] This is clearly reflected in a recent edited volume (Jasanoff and Kim 2015) where the concept of sociotechnical imaginaries is explored on a very broad canvas. The foci of the individual contributions cover a wide variety of topics, some of which deal with themes and dynamics that are related, if indirectly, to central concerns of civil defence history.[8] As an analytical concept, sociotechnical imaginaries appear to be catching on. An openness towards interdisciplinary scholarship and a theoretical and methodological approach that resonates in the contemporary human sciences are certainly part of the explanation. Substantially, the concept also appears to capture central features of modern existence. Sociotechnical imaginaries provide a gaze on the future that in part serves to structure it. Yet, they are rarely complete nor are they static, and they are often the subject of debate and contestation.[9]

In developing the concept, it has been an explicit ambition of Jasanoff to bridge two binaries that tend to stifle research in the human sciences. The ambition to overcome the first binary, that between ideas and materiality, is reflected in the very term *sociotechnical imaginary*, and this is, arguably, especially relevant for that large part of civil defence activities that contended with the threats posed by nuclear weapons technologies. After the bombings of Hiroshima and Nagasaki in 1945, few observers doubted the devastation that these weapons could produce; yet, at the same time, it became a peculiar quality of nuclear and later thermonuclear weapons that their subsequent use was restricted to testing and political signalling. Consequently, the threat of nuclear war was both vivid and real, and many of the defensive measures employed had a tangible material quality to them—yet the war itself unfolded in a future that was the province of the imagination (Grant and Ziemann 2016).

The second binary that this approach seeks to transcend, that between agency and structure, points towards analytical strategies that combine 'some of the subjective and psychological dimensions of agency with the structured hardness of technological systems, policy styles, organizational behaviours, and political cultures' (Jasanoff 2015a: 24). This ambition fits well with a marked trend in historical scholarship over recent decades: the rapprochement between cultural and political history where the category of 'politics' is expanded to encompass 'the political' and where power is conceived as multidimensional and not merely instrumental.[10] This ambition must be integral to any history of civil defence intent on venturing beyond personal testimony or organisational history. Even a brief glance at civil defence history highlights the difficulties of separating the political from the cultural—exercises and evacuation plans are obvious cases in point—and recent work is clearly in tune with these broader developments in historical scholarship (e.g. Cronqvist 2015). By foregrounding this ambition, however, Jasanoff encourages scholars to exploit a wide variety of source material that is available. In the case of civil defence, this includes official archives and policy documents, material artefacts and architecture, scientific work and philosophical reflection, official information and popular culture, as well as personal memoires and oral history.

Finally, Jasanoff has concluded the first sustained attempt to demonstrate the reach and insights of the concept by providing a heuristically useful structure for analysing the operation and life cycle(s) of sociotechnical imaginaries. This structure is divided into four phases. The first, *origins*, examines the mutual imbrication of scientific advances, technological developments, and social and political ideas in their historical context in order to determine how the contours of such imaginaries emerge. The second phase, *embedding*, is, broadly speaking, concerned with the institutionalisation of an imaginary whereby 'the merely imagined is converted into the solidity of identities and the durability of routines and things' (Jasanoff 2015b: 323). The third phase, *resistance*, refers on the one hand to the fact that large-scale reorganisations of human societies or their infrastructure typically involved slow and tardy processes. On the other hand, resistance points to the political or normative opposition that is a typical response to the formulation and institutionalisation of new imaginaries. What is resisted, however, can be anything from specific policies to visions of the future and their underlying social norms. The final phase, *extension*, places special emphasis on how imaginaries travel in time and space, often with the assistance of 'translation agents who are capable of moving imaginaries from one sociopolitical setting to another' (Jasanoff 2015b: 333).

Applying and Adjusting the Concept

The concept of sociotechnical imaginaries provides an instructive framework for exploring the hybridity of the ideational and the material that is characteristic of civil defence efforts as they developed during the Cold War. Apart from drawing attention to the intersection of science, technology and social life and facilitating transnational perspectives, there are several, more specific reasons for why the concept is productive and promising to think with in advancing civil defence history.

First, it is highly useful to place technology centre stage in approaching civil defence. After all, civil defence efforts were initially conceived as security measures against technologically mediated, man-made risks, above all forms of warfare that targeted civilian populations from the air by means of conventional explosives and later atomic bombs. In this early phase as well as in later developments, the logic and operation of civil defence were dependent on the employment of new scientific knowledge and a host of modern technologies (from technologies of governance and communication to those of underground construction and radiation detection) in an extensive and often overwhelming quest to ensure the survival of civil society. Science and technology were imbricated in this core rationale from the outset, since the foremost threats were products of weapons technologies based on scientific advances.

This circular logic and central (co-produced) predicament of civil defence became especially urgent and challenging in the nuclear age. It is a predicament, we contend, that comes into sharper focus by approaching civil defence history through the analytical perspective offered by a focus on sociotechnical imaginaries. The reference to desirable futures in the definition of the concept is likely to raise some eyebrows when the concept is applied to civil defence. Yet, notions of the desirable and the undesirable are hard to disentangle, and many of the agents involved in formulating, pursuing, practising or upholding civil defence were not blind to risks or problems but sought to manage, solve or contain them in ways that supported a desirable (or a least undesirable) future. Whether and how such imaginaries were supportive of or opposed to specific developments in science and technology, however, is best seen as an empirical question.

Second, the nature and history of civil defence efforts suggest that they contain a deep, perhaps even constitutive, ambivalence between fear and hope—an ambivalence that has also been found in wider studies of nuclear culture (Boyer 1985; Kinsella 2005; Hogg 2016; Sylvest 2020b). Significantly, however, while the hopes invested in civil defence activities may have had utopian qualities (see, e.g., Rose 2001), they just as often reflected a longing for maintaining or extending into a postwar situation the operation of existing norms and aspirations governing social life. The central place accorded to social norms in the study of sociotechnical imaginaries is highly beneficial in this respect. Whether the objective was to prepare for the worst or to discipline citizens (or both), the repeated imagination, rehearsal and contemplation of the next crisis, the next war and the postwar condition were central to civil defence.

The historical analyses in this book turn up many examples of how—within civil defence organisations, among civil defence personnel and in material civil defence structures—such dynamics translated into complex representations and constructions of technologies, often with a view to entrenching or shifting social norms (see also Clarke 1999). In this context, a key ambition invested in the concept of sociotechnical imaginaries is to illuminate how power relations and governance affect—and are affected by—visions of the role of technoscience in society. The exceptional shelter policy of Switzerland during the Cold War is an instructive case in point. As Berger Ziauddin (2017b: 676–677) has convincingly argued, Swiss nuclear bomb shelter policy can be viewed as a state-led attempt to install 'arrangements of protection and control into the private sphere' with a view to engineering 'docile, calm Cold War subjects'. This extensive quest to produce a 'bomb-proof' society was partly based on and served to extend existing national mindsets relating to the Swiss nation and the role of the family within it.[11] The ambivalence towards technology underlying this quest—channelled through vivid imaginations of fear and hope in malevolent and positive uses of technology—may even be constitutive of civil defence conceived as a sociotechnical imaginary.[12]

There is, then, ample potential in exploiting the concept of sociotechnical imaginaries for charting new courses in the field of civil defence history. On the other hand, the concept does not constitute a formula that can be applied to civil defence history without adjustments. While useful to think with, the analytical value of central dimensions of the concept can be developed and explored empirically using civil defence as a case. If done right, we suggest historical scholarship on civil defence can therefore both benefit from and contribute depth and texture to the ambitions that inform this approach. While the individual studies in this book display variation not only in how the concept is operationalised in relation to specific research questions but also in their reflections on the value of this perspective, the exploration of the concept and its analytical potential undertaken in this volume has been informed by a collective theoretical discussion. Two themes that warrant further elaboration or comment have stood out: the notion of life cycles and the question of scale.

First, while the conceptualisation of a 'life cycle' of imaginaries is heuristically useful, we think the phases identified by Jasanoff are most productively thought of as entangled and potentially reversible processes. In existing civil defence history, processes of (what could be seen as) embedding and resistance have been examined in some detail. It is not hard to see why. The sheer scale and complexity involved in designing, implementing and promoting a wide-ranging system like civil defence has long fascinated historians, just as the multiple forms of resistance such projects encountered—whether geographical, technological, financial, political, aesthetic or psychological—have attracted attention. Similarly, extension is clearly important for advancing the transnational and comparative dimensions of civil defence history, since it directs attention to the circulation, transfer and reception of ideas and artefacts that undergird civil defence conceived as an imaginary.

On the other hand, theoretically speaking it is far from clear whether and how embedding, resistance and extension follow one another or develop simultaneously. For example, embedding is absolutely central in studying civil defence, yet it is not clear at what point embedding takes on the character of extension. The category of origins, while logically predating the other phases and referring to an inescapable facet of historical scholarship, is deceptive. Indeed, to many historians it would be hubristic to straightforwardly claim to have identified the origins of any sociotechnical imaginary, as every apparent innovation is anchored in and remediates other structures, imaginaries and technologies. Several studies have pointed to important continuities between the Second World War and the early Cold War with respect to civil defence (Grossman 2001; Cronqvist 2012), while other studies have emphasised the importance of the interwar imagination of the next war for the development of defence practices during the Second World War (Bennesved and Norén 2020). In short, the linearity with which the notion of a life cycle and its various categories seem to be imbued risk installing a teleological bent in the concept of sociotechnical imaginaries. A great number of historical questions—for example about everyday experiences or memories of civil defence or the waxing and waning of the salience of nuclear war—are likely to require these processes to be collapsed and studied as deeply entangled, which in turn will disrupt lingering notions of sequence or 'progress'.[13]

Secondly, the question of scale presents a practical problem in operationalising the concept of sociotechnical imaginaries for the study of civil defence history. In Jasanoff and Kim's work, the original definition of the concept was national in scope, but for good reasons this has been abandoned. While it does make sense intuitively to speak of national civil defence policies and activities as constituting sociotechnical imaginaries, it may be possible to speak also of local, regional and transnational imaginaries, rival imaginaries or official and unofficial imaginaries just as it may be possible to conceive of forms of resistance as separate imaginaries (Farbøl 2021). In the same vein, imaginaries can be studied at various level of abstraction. Thus, whether civil defence is conceptualised as a separate imaginary in a particular historical context or whether it is best approached as forming part of or intersecting with a broader imaginary operative in a particular time and place (e.g. security politics, industrial modernity or a particular understanding of national purpose) will depend on a combination of empirical factors and the guiding scholarly interest.

In this context, it is also worth highlighting that several contributions to this volume also seek to attune the concept of sociotechnical imaginaries to one of the central challenges of civil defence history: to resist the temptation to study civil defence from the top down and instead to direct scholarship to how it was lived and practised on the ground, which was not always consistent with blueprint designs. Clearly, central dimensions of civil defence history are embedded in the decisions of policymakers and the documents and routines of organisations, but they are also produced by groups and individuals who through training exercises and everyday activities accorded the phenomenon alternative or additional meaning and purpose. In these practices, as they manifested themselves on different levels, we learn more about the 'collectively held, institutionally stabilized, and publicly performed visions' (Jasanoff 2015a: 4) of both peaceful and postwar futures. In the conclusion to this book, we offer further reflections on how the empirical analyses of this volume contribute to the approach.

Outline of the Book

The chapters in this book can be read as independent empirical analyses, but as a collective they provide valuable insights into transnational and translocal connections, similarities and differences in European histories of civil defence. The reader may identify an overall chronological movement from the early 1950s to the 2010s, but more importantly, there is a three-step thematic movement in the volume. The first two chapters deal predominantly with the *planning and structuring* of civil defence, the following four chapters largely centre on *exercising and communicating* civil defence, whereas the final two chapters are concerned with the *resisting and remembering* of civil defence.

In the following chapter 2, 'Order on Their Home Fronts: Imagining War and Social Control in 1950s NATO', Iben Bjørnsson connects the history of fear to civil defence efforts and analyses the discussions in NATO's Civil Defence Committee and Senior Civil Emergency Planning Committee following the introduction of the H bomb in the 1950s. By highlighting the relation between security politics and civil defence at a crucial period in the Cold War, the chapter is an ideal starting point for the volume as a whole. The main objective of NATO military and civil defence planners at this time was to secure the morale and obedience of populations by combining previous wartime experiences with scientifically grounded knowledge of the challenges of the thermonuclear age. In her analysis, Bjørnsson argues that the concepts of civil emergency planning and civil defence were closely intertwined with social control and hence also desirable social orders. The chapter also illustrates the importance of a transnational view of civil defence by paying attention to the coordination beyond national borders.

The planning of civil defence is also addressed in Chapter 3, 'The Imagined Disastrous: West German Civil Defence Between War Preparation and Emergency Management 1950–1990', by Jochen Molitor. The experiences of the Second World War were in no other country as complex as in the two Germanies. In the Federal Republic of Germany, the concept of civil defence appeared attractive after notions of 'total defence' had met with resistance and accompanying shelter construction programmes were discontinued. Officials who had initially spoken of 'disasters' to avoid the unpopular term 'war' began focusing on less obtrusive, attainable all-hazards measures such as volunteer relief work from the mid-1960s onward. Similarly, as Molitor shows, the majority of German physicians involved in disaster medicine pragmatically imagined scenarios up to, but not exceeding, a nuclear reactor meltdown or isolated nuclear strikes. In the wake of political détente and various technical disasters during the 1980s, most critics came to agree that at least some specialised preparation for more extreme scenarios might be warranted and did not necessarily amount to warmongering.

The following four chapters shift the focus from civil defence as a writing desk product to the different challenges facing the practitioners of civil defence on various levels and in different countries. Exercising and communicating civil defence brings to light the professional and everyday practices. Chapter 4, 'Normalising Nuclear War: Narrative Scenarios, Imaginative Geographies and Sites of Leisure in 1950s Britain' by Jonathan Hogg, directs our attention to the localised social, geographical and discursive contexts in which civil defence was anchored. There is a clear element of planning and structuring here, but since Hogg situates it on a local and not national or transnational level, the enactment and narrativisation of civil defence are at the centre of inquiry. The chapter brings up the narrative scenarios that were created to frame civil defence exercises, their public representation and also the sites of leisure and forms of civic engagement linked to civil defence activity.

Chapter 5, 'Embedding Preparedness, Assigning Responsibility: The Role of Film in Sociotechnical Imaginaries of Civil Defence', studies filmmaking in two neighbouring countries that differed both in terms of their political position in the Cold War and in the scale of their civil defence efforts: Sweden and Denmark. The comparison of the two countries finds a considerable degree of similarity in how films and their circulation were used to frame technologies and script and perform social norms during a period when the H bomb disrupted existing notions of civil protection. Zooming in on the functions of film in imaginaries of civil defence during the long 1950s, Bennesved and Sylvest argue that the difficulties of embedding sociotechnical imaginaries of civil defence in the early nuclear age are incapsulated in these filmic attempts to recruit volunteers and mobilise citizens for the cause of preparedness.

The thematic focus of Chapter 6 is the role of volunteering women in civil defence, and the scene here is the Netherlands. In '"The World is Her Home': The Role of Women Volunteers in Dutch Civil Defence in the 1950s and 1960s', Dick van Lente compares the efforts to set up and maintain a civil defence organisation with a campaign designed to prepare Dutch women for dealing with the consequences of a nuclear attack. Van Lente argues that the relative success of the women's organisation, compared to the poor performance of the male-dominated civil defence organisation, is best pursued in the context of changing views of the general role of women in Dutch society at the time. Indeed, van Lente identifies a conflict between the imaginary of civil defence advanced by Dutch civil defence authorities and the ideals and self-understanding of women volunteers, which was based on a rearticulation of social responsibility in the face of the Cold War.

Chapter 7, 'Ruins of Resilience: Imaginaries and Materiality Imagineered and Embedded in Civil Defence Architecture', brings up yet another area of civil defence practice, namely the material and technological history of civil defence training. Rosanna Farbøl examines a specific architectural phenomenon here, the so-called ruin town, that was built as a civil defence training ground. The phenomenon was present in several countries and Farbøl's particular case is Denmark. By analysing the ruin town as a stage for enacting and performing the anticipated war, she demonstrates how the ruins textualised and spatialised a sociotechnical imaginary that emphasised resilience and regeneration. The material powers of the ruins, in turn, standardised and homogenised the imaginary. In fact, Farbøl argues that ruin towns caused a 'taming' of the nuclear catastrophe and helped reinforce existing social norms and visions of the good society.

The last two chapters in this volume bring up different aspects of reacting to, resisting and remembering civil defence. In Chapter 8, 'Framing Civil Defence Critique: Swiss Physicians' Resistance to the Coordinated Medical Services in the 1980', Sibylle Marti brings up the so-called Coordinated Medical Services in Switzerland, whose task was to create prevention measures for a future nuclear war through an all-encompassing system of underground hospitals and first-aid stations. Marti analyses the criticism and resistance from different groups in the 1980s, and one in particular, Physicians for Social Responsibility (PSR), who refused to serve in the Coordinated Medical Services and in doing so gained much public and media attention. As Marti shows, by envisioning the impossibility and hopelessness of a life after a nuclear war, PSR challenged the authorities' sociotechnical imaginary with widely acclaimed alternatives.

Finally, in Chapter 9, 'Remembering Desirable Futures? Civil Defence Memories and Everyday Life', Marie Cronqvist and Matthew Grant explore to what extent the concept of sociotechnical imaginaries can be employed to examine civil defence as remembered. Using oral history sources from Sweden and the UK collected in the 2010s, they discuss the memory work of civil defence on three levels: localities, temporalities and mediations. Their analysis points to not only the ever-present and remaining narratives of resistance in both countries, but also the extent to which civil defence memories tend to diverge from the neat paths of any civil defence sociotechnical imaginary. It also shows that there is a certain 'fuzziness' of everyday life which problematises the binaries set out in the sociotechnical imaginaries framework regarding ideas/materiality and structure/agency.

The concluding chapter of the book provides reflections on the potential and limitations of the concept of sociotechnical imaginaries in historical scholarship. It also sums up the findings in the book as a whole by bringing forth the principal empirical similarities and differences that emerge from the chapters. Finally, we highlight the value of civil defence history for current imaginaries of security for European civil societies in the face of a highly diverse range of real and perceived threats.

Notes

1. Annex 6, Horizon Europe Programme (HORIZON), Draft Work Programme 2021–2022, 6. Cluster 3: Civil Security for Society, Version 16 October 2020, pp. 140ff.
2. See, for example, Hanson (2008), Baum et al. (2002), Jebari (2015), and Turchin and Green (2017). See also http://immortality-roadmap.com/sheltersmaps2.pdf (28 January 2021).
3. See Weart (1988, 2012), Whitfield (1991), Henriksen (1997), and Boyer (1998).
4. On USA civil defence history, see, for example, Brown (1988), Oakes (1994), McEnaney (2000), Rose (2001), Garrison (2006), and Davis (2007). On Germany, see Biess (2009), Diebel (2017); on the GDR, see Heitmann (2006). On the UK: Grant (2010), Douthwaite (2019). Canada: Burtch (2011). Sweden: Cronqvist (2012, 2015), Bennesved and Norén (2020), Bennesved (2020). Denmark: Sylvest (2018, 2020a), Farbøl (2020). Switzerland: Berger Ziauddin (2017a, b), and Berger Ziauddin and Marti (2020).
5. Recent work has ventured into some of these areas. See, for example, Masco (2008), Monteyne (2011), Grant (2019), Deville

et al. (2014), Douthwaite (2019), Berger Ziauddin (2017a), and Berger Ziauddin and Marti (2020). Cf. also Berger Ziauddin et al. (2018).

6. Recently, studies of urban civil defence have demonstrated the value of a localised perspective, see Singer (2015), Schregel (2015), and Farbøl (2020).
7. STS scholarship has played a major role in drawing attention to the social import of non-human entities and materialities, how anything from microbes to buildings matters in social life. This dimension of STS scholarship is primarily associated with Actor-Network Theory (ANT) and the work of Bruno Latour. See Latour (2005).
8. These include studies of US military science policy in the early Cold War (Dennis 2015) and the risk assessments underlying contemporary global health policy (Lakoff 2015).
9. See Sismondo (2020: 505). For examples of recent studies that deploy the concept to study airspaces and oceans as well as sustainable energy provision and the fourth industrial revolution, see Lawless (2020), Robinson (2020), Sovacool et al. (2020), and Schiølin (2020).
10. Especially in Germany, this form of history that places special emphasis on the role of communication in social life has been developed under the heading of 'new political history'. See Steinmetz and Haupt (2013).
11. The intimate relationship between civil defence, identity, statehood and citizenship in the nuclear age has been noted by a number of scholars (including Grant 2011; Cronqvist 2012; Deville et al. 2014; Sylvest 2018).
12. By retrieving visions, structures, practices and materialities of civil defence and exploring their relations to science and technologies, the concept may also carry potential for integrating civil defence with contemporary discussions of cultural heritage. See, for example, the discussion in Rindzevičiūtė (2019).
13. We thank Matthew Grant for raising this concern.

References

Baum, Seth D., David C. Denkenberger, and Jacob Haqq-Misra. 2002. Isolated refuges for surviving global catastrophes. *Futures* 72: 45–56.

Bennesved, Peter, and Fredrik Norén. 2020. Urban catastrophe and sheltered salvation. *Media History* 26 (2): 167–184.

Bennesved, Peter. 2020. *Sheltered society: Civilian air raid shelters in Sweden—From idea to materiality, 1918–1940 and beyond*. Malmö: Universus Academic Press.

Bennett, Luke. 2011. The bunker: Metaphor, materiality and management. *Culture and Organization* 17 (2): 155–173.

Bennett, Luke. 2018. Cold War ruralism. *Journal of Planning History* 8 (17): 205–225.

Berger Ziauddin, Silvia. 2017a. Superpower underground: Switzerland's rise to global bunker expertise in the atomic age. *Technology and Culture* 58 (4): 921–954.

Berger Ziauddin, Silvia. 2017b. (De)territorializing the home: The nuclear bomb shelter as a malleable site of passage. *Environment and Planning D: Society and Space* 35 (4): 674–693.

Berger Ziauddin, Silvia, and Sibylle Marti. 2020. Life after the bomb: Nuclear fear, science, and security politics in Switzerland in the 1980s. *Cold War History* 20 (1): 95–113.

Berger Ziauddin, Silvia, Peter Bennesved, and Sarah Robey. 2018. Living (again) in an age of nuclear fear: New avenues for studying Cold War civil defense, 16 April 2018. http://www.berlinerkolleg.com/sites/default/files/robey_berger_ziauddin_bennesved_cold_war_civil_defense_en.pdf [Accessed 11 May 2020].

Biess, Frank. 2009. 'Everybody has a chance': Nuclear angst, civil defence, and the history of emotions in Postwar West Germany. *German History* 27 (2): 215–243.

Boyer, Paul. 1985. *By the bomb's early light: American thought and culture at the dawn of the atomic age*. New York: Pantheon.

Boyer, Paul. 1998. *Fallout: A historian reflects on America's half-century encounter with nuclear weapons*. Columbus: Ohio State University Press.

Brown, Joanne. 1988. 'A is for atom, B is for bomb': Civil defense in American public education, 1948–1963. *Journal of American History* 75 (1): 68–90.

Burtch, Andrew. 2011. *Give me shelter: The failure of Canada's Cold War civil defense*. Vancouver: University of British Columbia Press.

Clarke, Lee. 1999. *Mission improbable: Using fantasy documents to tame disaster*. Chicago: Chicago University Press

Collier, Stephen J., and Andrew Lakoff. 2008. Distributed preparedness: The spatial logic of domestic security in the United States. *Environment and Planning D: Society and Space* 26: 7–28.

Cronqvist, Marie. 2012. Survival in the welfare cocoon: The culture of civil defence in Cold War Sweden. In *Cold War cultures: Perspectives on Eastern and Western European societies*, eds. Anette Vowinckel, Marcus M. Payk, and Thomas Lindenberger, 191–210. New York: Berghahn.

Cronqvist, Marie. 2015. Evacuation as welfare ritual: Cold War media and the Swedish culture of civil defense. In *Nordic Cold War cultures: Ideological promotion, public reception, and East–West interactions*, eds. Valur Ingimundarson and Rosa Magnusdottir, 75–95. Helsinki: Aleksanteri Cold War Series.

Davis, Tracy C. 2007. *Stages of emergency: Cold War nuclear civil defense.* Durham: Duke University Press.

Dennis, Michael Aaron. 2015. Our monsters, ourselves: Reimagining the problem of knowledge in Cold War America. In *Dreamscapes of modernity: Sociotechnical imaginaries and the fabrication of power*, eds. Sheila Jasanoff and Sang-Huyn Kim, 56–78. Chicago: Chicago University Press.

Deville, Joe, Michael Guggenheim, and Zuzana Hrdličková. 2014. Concrete governmentality: Shelters and the transformations of preparedness. *The Sociological Review* 62 (S1): 183–210.

Diebel, Martin. 2017. *Atomkrieg und andere Katastrophen.* Paderborn: Ferdinand Schöningh.

Douthwaite, Jessica. 2019. '… what in the hell is this?' Rehearsing nuclear war in Britain's civil defence corps. *Contemporary British History* 33 (2): 187–207.

Farbøl, Rosanna. 2020. Urban civil defence: Imagining, constructing and performing nuclear war in Aarhus. *Urban History*. https://doi.org/10.1017/S0963926820000590.

Farbøl, Rosanna. 2021. Prepare or resist? Cold War civil defence and imaginaries of nuclear war in Britain and Denmark. *Journal of Contemporary History*. https://doi.org/10.1177/00220094211031996

Fridlund, Mats. 2011. Buckets, bollards and bombs: Towards subject histories of technologies and terrors. *History and Technology* 27 (4): 391–416.

Garrison, Dee. 2006. *Bracing for Armageddon: Why civil defense never worked.* Oxford: Oxford University Press.

Geist, Edward M. 2019. *Armageddon insurance: Civil defense in the United States and Soviet Union, 1945–1991.* Chapel Hill: University of North Carolina Press.

Grant, Matthew. 2010. *After the bomb: Civil defence and nuclear war in Britain, 1945–1968.* Basingstoke: Palgrave Macmillan.

Grant, Matthew. 2011. Civil defence gives meaning to your leisure: Citizenship, participation, and cultural change in Cold War recruitment propaganda 1949–54. *Twentieth Century British History* 22 (1): 52–78.

Grant, Matthew. 2019. Making sense of nuclear war: Narratives of voluntary civil defence and the memory of Britain's Cold War. *Social History* 44 (2): 229–254.

Grant, Matthew, and Benjamin Ziemann, eds. 2016. *Understanding the imaginary war: Culture, thought and nuclear conflict, 1945–90*. Manchester: Manchester University Press.

Grossman, Andrew D. 2001. *Neither dead nor red: Civil defense and American political development during the early Cold War*. New York: Routledge.

Hanson, Robin. 2008. Catastrophe, social collapse and human extinction. In *Global catastrophic risks*, eds. Nick Bostrom, and Milan M. Cirkovic, 363–737. Oxford: Oxford University Press.

Heitmann, Clemens. 2006. *'Schützen und helfen'? Luftschutz und Zivilverteidigung in der DDR 1955 bis 1989/90*. Berlin: Christoph Links Verlag.

Hennessy, Peter. 2002. *The secret state: Whitehall and the Cold War*. London: Allen Lane.

Henriksen, Margot A. 1997. *Dr Strangelove's America: Society and culture in the atomic age*. Berkeley: University of California Press.

Hogg, Jonathan. 2016. *British nuclear culture: Official and unofficial narratives in the long 20th century*. London: Bloomsbury.

Jebari, Karim. 2015. Existential risks: Exploring a robust risk reduction strategy. *Science Engineering Ethics* 21: 541–555.

Jasanoff, Sheila. 2004. The idiom of co-production. In *States of knowledge: The co-production of science and social order*, ed. Sheila Jasanoff, 1–13. London: Routledge.

Jasanoff, Sheila. 2015a. Future imperfect: Science, technology and the imaginations of modernity. In *Dreamscapes of modernity: Sociotechnical imaginaries and the fabrication of power*, eds. Sheila Jasanoff and Sang-Huyn Kim, 1–33. Chicago: Chicago University Press.

Jasanoff, Sheila. 2015b. Imagined and invented worlds. In *Dreamscapes of modernity: Sociotechnical imaginaries and the fabrication of power*, eds. Sheila Jasanoff and Sang-Huyn Kim, 321–341. Chicago: Chicago University Press.

Jasanoff, Sheila, and Sang-Huyn Kim, eds. 2015. *Dreamscapes of modernity: Sociotechnical imaginaries and the fabrication of power*. Chicago: Chicago University Press.

Karl, Krystin, and Alex Wellerstein. 2017. About the project. https://reinventingcivildefense.org/project [Accessed 3 January 2018].

Karl, Krystin, and Ashley Lytle. 2019. This is not a drill: Lessons from the false Hawaiian missile alert. *Bulletin of Atomic Scientists*. https://thebulletin.org/2019/01/this-is-not-a-drill-lessons-from-the-false-hawaiian-missile-alert [Accessed 11 January 2019].

Kinsella, William J. 2005. One hundred years of nuclear discourse: Four master themes and their implication for environmental communication. *Environmental Communication Yearbook* 2 (1): 49–72.

Lakoff, Andrew. 2007. Preparing for the next emergency. *Public Culture* 19 (2): 247-271.

Lakoff, Andrew. 2015. Global health security and the pathogenic imaginary. In *Dreamscapes of modernity: Sociotechnical imaginaries and the fabrication of power*, eds. Sheila Jasanoff and Sang-Huyn Kim, 300–320. Chicago: Chicago University Press.

Latour, Bruno. 2005. *Reassembling the social: An introduction to actor-network theory*. Oxford: Oxford University Press.

Lawless, Christopher. 2020. Assembling airspace: The single European sky and contested transnationalities of European air traffic management. *Social Studies of Science* 50 (4): 680–704.

Masco, Joseph P. 2008. 'Survival is your business': Engineering ruins and affect in nuclear America. *Cultural Anthropology* 23 (2): 361–398.

Masco, Joseph P. 2013. From *The nuclear borderlands* to the counter-terrorist state: An interview with Joseph P. Masco. *Social Anthropology* 21: 389–402.

May, Elaine Tyler. 1988. *Homeward bound: American families in the Cold War era*. New York: Basic Books.

McEnaney, Laura. 2000. *Civil defense begins at home: Militarization meets everyday life in the fifties*. Princeton: Princeton University Press.

Molitor, Jochen. 2011. *Mit der Bombe überleben: Die Zivilschutzliteratur der Bundesrepublik 1960–1964*. Marburg: Tectum Verlag.

Monteyne, David. 2011. *Fallout shelter: Designing for civil defense in the Cold War*. Minneapolis: University of Minnesota Press.

MSB (Myndigheten för samhällsskydd och beredskap). 2018. *Om krisen eller kriget kommer* [If crisis or war comes]. Karlstad: Swedish Civil Contingencies Agency. Available at: https://www.msb.se/RibData/Filer/pdf/28494.pdf [Accessed 11 May 2020].

Oakes, Guy. 1994. *Civil defense and American Cold War culture*. Oxford: Oxford University Press.

Rindzevičiūtė, Egle, ed. 2019. *Nuclear cultural heritage position paper*. HRC Research Networking Project, AH/S001301/1. Kingston upon Thames: Kingston University. Available at: https://nuclearculturalheritage.files.wordpress.com/2019/11/2019-nuclear-cultural-heritage-position-statement.pdf [Accessed 28 January 2021].

Robinson, Sam. 2020. Scientific imaginaries and science diplomacy: The case of ocean exploitation, *Centaurus*. https://doi.org/10.1111/1600-0498.12342, 1–21.

Rose, Kenneth D. 2001. *One nation underground: The fallout shelter in American culture*. New York: New York University Press.

Schiølin, Kasper. 2020. Revolutionary dreams: Future essentialism and the sociotechnical imaginary of the fourth industrial revolution in Denmark. *Social Studies of Science* 50 (4): 542–566.
Schregel, Suzanne. 2015. Nuclear war and the city: Perspectives on municipal interventions in defence (Great Britain, New Zealand, West Germany, USA, 1980–1985). *Urban History* 42 (4): 564–583.
Singer, Eric. 2015. Civil defence in the city: Federal policy meets local resistance in Baltimore, 1957–1964. *Urban History* 42 (4): 547–563.
Sismondi, Sergio. 2020. Sociotechnical imaginaries: An accidental themed issue. *Social Studies of Science* 50 (4): 505–507.
Sovacool, Benjamin K., et al. 2020. Imagining sustainable energy and mobility transitions: Valence, temporality, and radicalism in 38 visions of a low-carbon future. *Social Studies of Science* 50 (4): 642–679.
Stafford, James. 2011. 'Stay at home': The politics of nuclear civil defence, 1968–83. *Twentieth Century British History* 23 (3): 383–407.
Steinmetz, Willibald, and Heinz-Gerhard Haupt. 2013. The political as communicative space in history: The Bielefeld approach. In *Writing political history today*, eds. Willibald Steinmetz, Ingrid Gilcher-Holtey, and Heinz-Gerhard Haupt, 11–43. Frankfurt: Campus.
Sylvest, Casper. 2018. Atomfrygten og civilforsvaret. *temp—tidsskrift for historie* 16: 16–39.
Sylvest, Casper. 2020a. Nuclear fallout as risk: Denmark and the thermonuclear revolution. In *Histories of knowledge in postwar Scandinavia: Actors, arenas and aspirations*, eds. Johan Östling, Niklas Olsen and David Larsson Heidenblad, 21–38. London: Routledge.
Sylvest, Casper. 2020b. Conceptions of the bomb in the early nuclear age. In *Non-nuclear peace: The Ban Treaty and beyond*, eds. Tom Sauer, Jorg Kustermans, and Barbara Segaert, 11–37. London: Palgrave Macmillan.
Turchin, Alexey, and Brian Turner Green. 2017. Aquatic refuges for surviving a global catastrophe. *Futures* 89: 26–37.
Vale, Lawrence J. 1987. *The limits of civil defence in the USA, Switzerland, Britain and the Soviet Union: The evolution of policies since 1945*. London: Palgrave Macmillan.
Weart, Spencer. 1988. *Nuclear fear: A history of images*. Cambridge, MA: Harvard University Press.
Weart, Spencer. 2012. *The rise of nuclear fear*. Cambridge, MA: Harvard University Press.
Whitfield, Stephen J. 1991. *The culture of the Cold War*. Baltimore: The John Hopkins University Press.

CHAPTER 2

Order on Their Home Fronts: Imagining War and Social Control in 1950s NATO

Iben Bjørnsson

'Keep Calm and Carry On'. The iconic UK wartime poster is one of many examples of governments instructing people how to behave in war or crisis. From *Duck and Cover* cartoons, to farmers' handbooks, civil populations grew used to seeing such instructions in the twentieth century. In this chapter, I ask what ideas they built on. What were the origins and expressions of those ideas, their traces in modern history and their development with (thermo)nuclear weapons?

The subject of analysis is the NATO Civil Defence Committee and the Senior Committee for Civil Emergency Planning. Being fora for exchange and (some degree of) coordination between Western countries, these committees are obvious but understudied sources when dealing with civil defence cultures and attitudes. They will be analysed in a broader framework of Western scientific culture, total war and, not least, the theoretical framework of the sociotechnical imaginary.

I. Bjørnsson (✉)
Stevnsfort Cold War Museum, Roedvig, Denmark
e-mail: ib@oesm.dk

M. Cronqvist et al. (eds.), *Cold War Civil Defence in Western Europe*,
https://doi.org/10.1007/978-3-030-84281-9_2

The time frame is the 1950s, which established the thermonuclear age. It is also NATO's first decade of existence in which it negotiated its role in the new world order.

The article falls in four sections. First, I introduce the theoretical framework of the study, as well as the historical background: the modern Western cultural tradition of dealing with human fear in times of crises. Following this I investigate how war and civil populations were imagined in NATO in the 1950s and, in the next section, how this was underpinned by science. Lastly, I ask what NATO thought could or should be done. In the conclusion, I will tie the analysis to the theoretical framework, to say something about the larger picture of what preparedness meant to NATO and Western society.

Sociotechnical Imaginaries and Concepts of War

The notion of the sociotechnical imaginary rests on the idiom of co-production, which holds that development is powered by intertwined processes of technology and science informing culture and politics, and vice versa. Sociotechnical imaginaries are such negotiations of science into power structures. Hence, they are normative: they express (un)desired orders and strive to have a governing and/or disciplining function; they too are instructions in behaviour (Jasanoff 2004: ch. 1–2; Jasanoff 2015: 2–4).

The definition at first glance seems to favour the positive, 'desirable futures'. But Jasanoff notes that a desirable outcome correlates with an obverse scenario: 'shared fears of harm' and 'negative imaginings'—utopia vs. dystopia (Jasanoff 2015: 4–5). She does actually extend the definition, albeit in a parenthesis, beyond 'collectively held and performed visions of desirable futures' to 'resistance against the undesirable' (Jasanoff 2015: 19). With this subtle addition, a sociotechnical imaginary can also be a safeguard against an undesired future or dystopia. As we shall see, this is the case in this chapter.

Imaginaries are constituted by '[p]erformances of statehood [...] tied to demonstrations and to public proofs employing scientific and technological instruments' (Jasanoff 2015: 9–11). Most obvious is the scientific test, but such performances can also take the form of moon landings, complicated surgery—or civil defence exercises: calm, rational behaviour acted out as both rehearsal and theatre, aimed to ensure 'calm and composed [...] subjects' (Ziauddin 2017: 9; Cronqvist 2008; Davis 2007).

In this chapter, I will operationalise the framework of sociotechnical imaginaries to look at how science underpinned social order(s) and what this can tell us about what was (un)desirable. For this purpose, it is important to note that the sciences employed might not only be technological, but also human and social.

One of the most common human reactions to war or catastrophe (and indeed danger at large) is fear. But while fear is among the most basic of instincts, historian Joanna Bourke shows how fear as a societal phenomenon changed character in the nineteenth and twentieth centuries. Scientific developments created new things to fear, and rising secularisation of Western societies shifted responsibility for handling societal fears from the church to secular authorities (Bourke 2006: 23–24). These authorities sought help from the burgeoning psychological and social sciences around the turn of the century. Hence, these sciences came to serve as a tool not only for better understanding the human mind, but also for defining desirable states of mind and treating undesirable ones. They were employed by authorities to make sense of modernity, but also advanced by modernity itself.

One manifestation of fear is panic. And a phenomenon of urban living is so-called mass panic. That this could be deadly became clear in the late nineteenth and early twentieth centuries when it broke out in enclosed areas, like theatres and on factory floors. Thus, panic was now something to be handled by authorities, and the regulation of public space for the purpose of (emotional) management of people became a marker of modernity (Bourke 2006: 51ff). Panic also became an area of research for the psychological sciences. In the late nineteenth century, theories of 'the group mind' identified panic as a contagious phenomenon, causing otherwise sensible and rule-abiding individuals to behave irresponsibly and immorally. Panic was the opposite of civilisation and sociability, efficiency and progress. The volatile crowd, uncontrolled, could wreak political, social and economic havoc. Sociologists spoke of disorganised and selfish behaviour as an expression of instinct normally covered in the veneer of civilisation, but susceptible to activation by fear (Bourke 2006: 61ff). Human fear, if not handled, could be a steppingstone towards chaos.

With the great wars of the twentieth century, the insights from these scientific delvings into human fear were put into practice on soldiers. A finding from the First World War was that fear could be calmed by, or channelled into, activity. The worst thing for the mental well-being of a soldier was being unable to do anything—the best was to be prepared and trained. Fear took nourishment from rumours and insecurity and could be relieved by the presence of calm and confident authority figures (Bourke 2006: 208ff; Davis 2007: 115). These observations were confirmed during the Second World War, in which the view also evolved that fear was not something one should strive to eradicate or be free of, but rather a natural and even healthy reaction to a critical situation which could spur further action (Biess 2009: 219–220; Bourke 2006: 212; Oakes 1994: 50, 62–64).

These studies, which initially looked at soldiers, were replicated in civilian life. In 1939, the British War Office conducted investigations into avoiding civil panic, the results of which were in line with what had already been established: fear and panic could be relieved by proper preparation, information control and a sense of responsibility. The ability to *do something* was essential, and this should be trained beforehand, as recognisability and repetition promoted a sense of security. If this did not work, however, there were other, hands-on methods: British civil defence personnel were trained in 1940 to 'treat' panic-stricken persons with a bucket of water or a punch in the nose or ear. Panic was contagious and had to be shut down immediately (Bourke 2006: 224, 244ff).

Studies of civilian populations' actions and their role in a war was a result of the totalisation of war, another feature of the twentieth century. This indicates that war was no longer just a military matter, but also a civil one. Total war was facilitated by technological advances, especially in air force technology, which led wars far into territories beyond the front. It was also helped along by radio and (moving) images bringing faraway battlefields into theatres and living rooms. Hence, governments sought to mobilise the *home front*, where maintaining the *morale* of civil populations was essential. Civilians were rallied to participate in the war effort through a steady stream of messages: grow vegetables, 'keep mum', buy war bonds. Keep calm, carry on. Civil defence is in itself an expression of the totalisation of war, which civil society was now asked to confront and tackle (McEnaney 2000; see also Hogg, this volume, Fig. 2.1).

In Cold War USA, civil panic was deemed so decisive for a war that it could even be used deliberately. The US Joint Chiefs of Staff believed the 'chaos and extreme confusion', 'hopelessness and shock' and 'aimless, even hysterical activity' following a nuclear attack on the enemy could work effectively to their own advantage (Oakes 1994: 33, 35–36).

Fig. 2.1 The mobile civil defence exhibition 'Alert America' travelled the USA in 1952–55
It was one of many attempts to mobilise the 'home front' and make preparedness for war a feature of civil society (Photo: US National Archives, https://www.digitalcommonwealth.org/search/commonwealth:h128rr89n)

In 1951, the US Federal Civil Defense Administration (FCDA) and the Pentagon launched a civil defence research project, *Project East River*, that repeated earlier conclusions: panic could lead to disobedience and even resistance, protests, vandalism and riots. The road from panic to civil unrest to uprisings seemed all too short, and panic was viewed as a subversive, perhaps even revolutionary, phenomenon. But project East River also found remedy in the established methods: preparation and having something to do combined with establishment of authority and information control (Oakes 1994: 40–42, 62ff; Grossman 2001: 59–61; Bourke 2006: 270ff; McEnaney 2000: 31–33; Davis 2007: 120).

Project East River's results became authoritative within American civil defence. It stood on the shoulders of existing war and disaster research, both American and European, and the results and recommendations were fundamentally the same. As such, East River both drew from existing research trends and fed back into them.

With Project East River in mind, it is perhaps not strange that the USA took a keen and early interest in panic prevention. President Eisenhower believed that, in war, both civilians and officials would lose their minds, and he was generally pessimistic about being able to do anything about it. According to Val Peterson, US Federal Civil Defense Director from 1953 to 1957, 'mass panic may be far more devastating than the bomb itself'.[1] One FCDA official estimated that the larger part of their work in the 1950s was about panic prevention (Oakes 1994: 41–42, 150–151; Davis 2007: 122; Bourke 2006: 268–269).

Another expression of the interest in (unwanted) mental reactions to disaster, which began in military medicine and worked its way into civil defence circles in the 1950s, was the development of psychological first aid as a field—the skill of soothing a shocked mind (Blain et al. 1945; Drayer et al. 1954).

We see here the contours of a sociotechnical imaginary in which panic and disorderly, individualised behaviour was undesirable. Its precise origins are hard to pinpoint, but modernity, urban masses and modern warfare were all elements in its creation. Formulated positively, this imaginary envisioned calmness in catastrophe fostered by proper training and education and adherence to authority: a *civil defence culture* of preparedness and deference (Cronqvist 2012). However, it was mostly formulated negatively, in terms of what would happen if preparedness failed: as an undesirable.

According to Lord Ismay, NATO General Secretary from 1952 to 1957, when he took office, he asked Supreme Allied Commander in Europe (SACEUR) General (later US President) Eisenhower how he, Ismay, could best be of service. Eisenhower answered that he needed Ismay to keep up morale on the home fronts.[2] In a memorandum to the North Atlantic Council shortly thereafter, Ismay wrote that modern war was no longer just a military effort, but 'a trial of strength and willpower between whole nations or groups of nations', and without a stable home front, the military effort would be futile.[3] The NATO Civil Defence Committee (CDC) were formed around the same time, in the summer of 1952. Civil defence was a direct result of the totality of modern war.

Imagining War and Civil Defence in 1950s NATO

When imagining war, people usually go by experience. This was also the case in postwar Europe (Grant 2016: 96). However, the Cold War contained a threat with which very few people had experience, namely nuclear weapons. The only people who had actually experienced their impact were in Japan. But while images were available to European audiences, this did not fundamentally alter the European vision of war: a bombed-out city looks the same whether its name is Hiroshima or Hamburg (Ziemann 2016: 117–118; Molitor and Hogg, this volume). Thus, Hiroshima and Nagasaki were woven into the existing European visions of war without fundamentally altering them.

Not having a recent memory of war on their own territory, Americans had to go 'on imagination alone' (Oakes 1994: 5). According to UK planners, this caused American authorities to focus more on panic, whereas British authorities considered their population already well trained in handling war (Grant 2010: 11). Or perhaps the Americans just took seriously their own research from Hiroshima and Nagasaki, which stated that:

> ... the primary reaction to the bomb was fear – uncontrolled terror, strengthened by the sheer horror of the destruction and suffering witnessed and experienced by the survivors. (...)
>
> The two typical impulses were those: aimless, even hysterical activity, or flight from the city to shelter and food.

An interviewee was quoted: 'I became hysterical (...) and we just ran around without knowing what to do'.[4]

Apart from these differences, both US and Western European authorities and (civil) defence planners initially thought a Third World War would be survivable. A war was not thought to be wholly nuclear, and the atomic bomb was mostly considered a very large, but otherwise normal bomb. Civil defence could build directly on the Second World War experience and did not have to undergo fundamental change. Oakes considers this 'conventionalisation' of the A bomb a stunt to reassure the population (Oakes 1994: 51ff), but the general impression from at least Danish and NATO sources is that civil defence authorities sincerely believed this. Whether that was just a case of them being duped by the Americans is not to be speculated on here. But it should be added that all discussions about nuclear weapons in the CDC were informed by Hiroshima/Nagasaki and nuclear tests made by the USA. As such, the American interpretation held sway.[5]

At a CDC meeting in late 1953, the American delegation showed the nuclear test film *Operation Doorstep*. It showed the effect of the A bomb on such things as wooden houses, subterranean shelters and cars. The blast was 15 kilotonnes (kt.)—Hiroshima-sized—and the comforting message was that less than a mile from the explosion itself (ground zero) humans could survive if properly prepared and sheltered. What it did not say, but which the American delegate disclosed at the meeting, was that 15 kt was already considered relatively small and that this bomb had been detonated at an altitude which minimised its effect. A larger bomb detonated higher in the air would have been 'much more devastating'.[6]

In May 1955, the FCDA invited the entire CDC to visit the Nevada desert themselves and observe another test, *Operation Cue*. This test was also framed reassuringly: although the reinforced wooden house 5,500 feet from the explosion lost roof, chimney and windows, the mannequins serving as the family in the basement were unharmed. The model power plant and communication lines also survived. Once again, the conclusion was that with proper preparation, an atomic explosion could be both managed and survived.[7] Footage of these tests were broadcast to the American public and are a prime example of the publicly performed scientific test which sought to attest to and solidify the civil defence imaginary (Fig. 2.2).[8]

However, in the early- to mid-1950s a new type of nuclear weapon emerged: the hydrogen (H) bomb. The strength of the H bomb is not counted in kilo- but in megatonnes (a thousand times more powerful), which made it increasingly impossible to pretend—or believe—that nuclear weapons could be likened to conventional bombs. The H bomb was accompanied by a shift in US and NATO strategic doctrines; the 1953 *New Look* strategy of the Eisenhower administration was based primarily on nuclear weapons and was followed by the NATO doctrine of *massive retaliation*—all-out nuclear response to an attack. Not only did nuclear weapons become more terrifying, but a future war became

Fig. 2.2 The nuclear test as a public performance and media event
FCDA Director Val Petersen is interviewed at the observer area of the Operation Cue atomic blast in 1955 (Photo: US National Archives, https://www.digitalcommonwealth.org/search/commonwealth:h128s189g)

almost wholly nuclearised (McMahon 2010: 293). As these developments made war equally unpredictable and indeed unimaginable on both sides of the Atlantic, experience no longer counted (Rose 2001: 38; Grant 2016: 101ff, 132; Horn 2016: 33–34, 47). British authorities started doubting that their otherwise experienced population would 'keep calm and carry on' in thermonuclear war. Panic ceased to be mainly an American concern (Grant 2010: 50, 112; Hennessy 2010: 153–154, 177).

Initially, the FCDA did try to maintain an image of the H bomb as merely a VERY big, but otherwise still normal, bomb (Oakes 1994: 58–62). This was evident at a May 1954 CDC meeting where FCDA Deputy Director Katherine Howard stated that its destructive effects were greatly exaggerated by the media. Though more powerful, it was not essentially different from the A bomb, so nothing would fundamentally have to change. She concluded that 'precautions could be taken against the effects of this weapon' and showed 'no sign of pessimism on this point'.[9]

This time, however, the rest of the committee was not convinced. During 1954, the Danish Civil Defence Directorate's scientists began questioning assumptions about protection against the H bomb.[10] This reached Director Arthur Dahl during his tour of the USA in autumn 1954, where he saw for himself that American civil defence planners—contrary to Howard's statements—were adjusting to thermonuclear reality by transitioning from a strategy of shelters to evacuation.[11] Upon return, he brought it up in the CDC, stating that many of the known protective measures were already outdated and that civil defence also needed a 'New Look'. British representative General F. Irving agreed. He stated directly that they had to stop talking about what had worked in the last war and start talking about what to do in a future one. The UK, he said, had no intention of spending more money on measures that did not also protect against thermonuclear weapons. Val Peterson admitted that the best defence against the H bomb was dispersal of people and vital installations, and he confirmed that the USA had started making extensive evacuation plans. CDC chairman Laborie asked for a revision of current work, and NATO Civil Defence Advisor John Hodsoll promised to do a review.[12] He was not the only one: centrally, NATO had also started work on new planning assumptions, which Hodsoll drew upon in preparing his report.

Hodsoll's report (ready in just a week) stated that the H bomb fundamentally altered things and that the 'general arrangement' of the last war was no longer sufficient. Where the goal then had been to uphold everyday life and production albeit at a reduced level, everything now had to be about immediate survival. 'Business as usual' did not exist when everything within a radius of seven miles from ground zero was devastated. Now, 'everything would have to be temporarily subordinated to the maintenance of life and health; in other words, to the basic essentials of human existence'. Invoking the language of total war, Hodsoll stated that measures would be 'aimed at preserving the civil population and sustaining their morale, since if they collapse their country will cease to be a combat unit' and concluded: 'The Home Front is fundamental. If it collapses all is lost'.[13]

The new NATO planning assumptions, which had informed Hodsoll's memorandum, were adopted by the North Atlantic Council in May 1955.

They heralded a new era, foreseeing that 'on the outbreak of hostilities, the enemy will launch thermo-nuclear attacks with first priority' and that 'thermo-nuclear weapons can create destruction and chaos on a scale exceeding anything hitherto contemplated'.[14] It was thought that an initial intense nuclear bombing period of three to seven days would be followed by a 30-day period which determined whether or not a nation survived. In this light, the Council encouraged NATO members to prioritise planning for survival in the first 30 days of war. Survival was defined as 'maintenance of the authority of the government and the preservation of human life'.[15]

These changes led to a reorganisation in which all NATO committees on civil planning, including the CDC, were put under a new Senior Committee for Civil Emergency Planning, for the purpose of securing cohesive and focused planning. This included but also went beyond civil defence: civil emergency planning encompassed all civil and societal measures directed at securing the survival of the nation in accordance with the Council's definition, including safe locations for governments, emergency laws, communication lines, infrastructure and shipping.[16]

Bringing in Science

The preoccupation with panic and upholding authority spread across the Atlantic and took a hold in NATO circles, even its top tier: in 1955, Deputy Supreme Allied Commander of Europe (DSACEUR) Field Marshall Montgomery stated that one of nuclear war's big dangers—which might even end up costing victory—was 'hysteria'.[17]

It also became the subject of interest for civil defence and preparedness sciences which, up to that point, had mainly dealt in more quantifiable topics such as thickness of bunker walls, blast effects and radiation levels. A sociotechnical imaginary is underpinned by, and intertwined with, science. Therefore, the emergence of psychology in scientific debates in civil defence circles also tells us something about how the imaginary developed. The increased focus on psychological sciences in NATO/Western European civil defence circles in the 1950s informs us of a sociotechnical imaginary that increasingly explicated its visions of behaviour.

By comparing and contrasting four articles by medical professionals on the subject in the time span of 1956–1958, we shall see that, although the thermonuclear era was new, the particular problem of panic and unrest, as well as its cure, were closely in line with previous thinking.

The articles are 'Views on psychological reactions in war' ('Synspunkter på psykiske reaktioner under krig') by Swedish Civil Defence Surgeon General Dr Walo von Greyerz, 'Maintenance of morale among civilians during a nuclear emergency' by Chairman of NATO's Medical Committee Dr Svend Toftemark of the Danish Ministry of Health, 'Civil morale under the threat of nuclear attack' by psychiatrist at the Department of Psychological Medicine at London University College Hospital Dr Roger Tredgold and 'Psychological considerations in Civil Defence Planning' by British professor of psychology Thomas Ferguson Rodger. All four articles struck the following themes in more or less detail: desired and undesired symptoms of fear and remedies in the form of preparation, information, activity, authority and—in some cases—force.

Some were very particular about describing symptoms. Von Greyerz divided fearful reactions into two categories, each containing three subcategories with symptoms ranked from mild to severe. The first category contained uneasiness, apprehension and anxiety. Uneasiness was a normal reaction to war, but via apprehension it could escalate to anxiety, which demanded treatment. Whereas the first category was permanent or longer-lasting problems, the symptoms in the second category—fear, dread and terror—were more immediate reactions. Otherwise the principle was the same—a normal reaction could escalate into an unhealthy one:

> During an air raid, we share fear with everyone around us, and we don't have to be ashamed of it. We must receive it as something natural. But if fear – via the intermediate stage, which I call dread – increases to terror, it will have reached proportions where it can no longer be accepted as an ally.[18]

If fear was allowed to escalate into terror there were two possible outcomes: apathy, where people would lose all ability to act and 'the storm', where they would run around bewildered, incapable of reasonable action, perhaps even displaying symptoms such as blindness, deafness, numbness or unconsciousness (a state which he labelled 'hysteria'). If a 'terror psychosis' 'infected' the surroundings, it could 'drive a whole group to actions that go against common sense; this is when panic arises'.[19]

Tredgold ranked reactions to fear more simply: as either 'healthy' or 'unhealthy'. Unhealthy reactions were 'panic, paralysis, overactivity, bodily complaints (based largely on anxiety) and depression which may lead to fatalistic (and fatal) resignation'. Fear in itself, however, was natural and only 'exaggerated' reactions to it were unhealthy.[20] Much the same was expressed by Toftemark and Ferguson Rodger. The latter, however, ranked panic as more detrimental than apathy. He stated that a 'dazed apathetic state' was a relatively normal reaction to catastrophe, and preferable to 'excited behaviour'.[21]

There were different views as to what panic looked like. To Tredgold, panic (or 'overactivity') might on the surface look purposeful, but on inspection it was more like a fish on dry land, 'floundering about to no purpose, gasping and with an occasional convulsive jerk'.[22] Ferguson Rodger took almost the opposite approach: purposeful behaviour could look like panic for the uninitiated, who only saw 'what appears to be random, uncontrolled activity' with people 'running or driving vehicles in opposite directions, often at times passing each other without acknowledgment or seeming awareness'. This, however, was 'not panic, but social disorganization. Many individuals and small groups are working within the disaster area with purpose and some degree of organization'. Nor was flight necessarily panicked; in fact, 'it is more often orderly and controlled', but could also, to the uninitiated look disorganised—which was why journalists often exaggerated and over-reported panic. He reminded the reader that panic had actually been rare during the last war.[23] This point was also made by von Greyerz, who reassured the reader that panic had 'only in exceptional cases been a serious problem'. He also made the point that what could look like panic at first glance, like mass flight, could in actuality be rational and 'have an objective in sight'.[24]

As such, they differed on how reactions should be categorised and on the nature of events of panic. But they all thought it best to prevent them altogether, and largely agreed on the means: 'The three big musts of panic prevention are preparation, information and action'—these were the words of Toftemark, although he was himself rather pessimistic as to how well it would work. He thought that even prior education of people could not ensure that they would react correctly in an emergency. Its effectiveness required them to be both psychologically and emotionally involved, which in turn required authorities to counter indifference not only with information, but also with positive motivation and a realistic picture of the options.[25] Von Greyerz also expressed the view that peacetime preparation and education was of the utmost importance—including the knowledge that fear was natural and could be remedied.[26] Tredgold made the same point, noting that what a threat meant to the individual 'will depend largely on what he has been told about it'.[27]

Ferguson Rodger thought that a prepared population would react adaptively and that the civil population could be trusted with the initiative in an emergency—the most important thing was that they be properly informed both ahead of but also during a catastrophe.[28] So everyone agreed that preparation was key, even if Toftemark was somewhat pessimistic about people's involvement in the prior preparation. This problem was, however, a known one. Civil defence authorities hit this wall time and again trying to 'educate' a public which seemed indifferent to or even ridiculed the attempts. However, the stance of the authorities was clear: it was better to repeat the message and hope something stuck than to have an utterly oblivious population (Davis 2007: 170).

Everyone also agreed that information to the population once disaster had struck was as crucial as information beforehand. Information could itself relieve fear: for instance, noise from anti-aircraft weapons would be perceived as less unnerving if people knew that it was a welcome part of their own defence. As von Greyerz stated: 'People who know what has happened are considerably better equipped to bear the burden of reality, than those who are in the dark'.[29] Ferguson Rodger recommended giving people 'clear and unambiguous information, even if it is of the gravest import'. If people were 'as well informed as possible', then they would stay rational and make the right choices. Lack of information, on the other hand, was a source of panic.[30] This also implied that even bad news was better than none. The unknown was always scarier than reality. According to Tredgold, there was a real chance of 'panic caused by gross over-imagination. I can understand that there may be reluctance to tell people horrible facts, but they will indulge in more horrible fears if they are not told'.[31]

This information, obviously, should come from people in authority, and this was another common feature of the articles. Von Greyerz stated that fear in a person could be relieved alone by someone else's 'firm demeanour',[32] and Toftemark emphasised the necessity of constructive and authoritative role models to prevent the spread of uncontrolled and individualised behaviour. While some initial 'disorganization' could not be avoided, the trick was to quickly establish organisation and action by strong leadership.[33] Ferguson Rodger's biggest worry was people isolated in fallout shelters for too long. This was not only bad for mental health in general but could also lead to the very lack of information which incited fear—and panic. Therefore, fallout shelters should be regularly patrolled by authorities with tidings.[34]

Authorities could also give people something to do, and this was another most relished remedy of fear with all authors. Fear and action were intertwined: fear could spur action and action remedied fear. Activity of any kind was good—according to von Greyerz, 'preferably a simple task which is physically strenuous, like handing water or carrying bricks'.[35] It could not only hinder fear from turning into panic, but also, according to Dr Threadgold, even remedy paralysis and 'bodily complaints'.[36]

Lastly, the authors mentioned those instances where the sufferer was not able to act on their own or under instruction because they had entered a state of panic. Both Toftemark and Tredgold mentioned the relatively new term 'psychological first aid' as a remedy that should be trained in advance. Tredgold recommended that people be led away from the disaster and calmed down before further action could be taken. Calming down could mean anything from a cup of tea to sedation, and the article focused on treating people with empathy, praising them when they did something sensible or worthwhile. If people turned to religion, this should be encouraged (whatever one's own personal attitude), as it could soothe mental suffering. Through these remedies, the doctor thought morale could be raised considerably, and mental damage prevented from being long-term. Von Greyerz also recommended sedation as treatment for both apathy and terror. Furthermore, he recommended removing the sufferer from others so as it would not 'rub off'. If this was not a possibility, for instance in a shelter, a last resort could be to tie down and gag the afflicted person, although not without taking care to explain to others present why such extreme measures were necessary.[37]

Apart from differences in nuances and categorisation, it is striking how similar these articles are, in both their diagnoses and recommended remedies. It is fair to say that Western civil defence psychology was a research field marked by consensus, and it is also worth noting that it correlates with earlier military psychology. Although it was a new field of research, it was rooted in a firm tradition (as was civil defence and disaster medicine as a whole, see Molitor, this volume). This sociotechnical imaginary that favoured calmness and preparedness had been extended from military medicine to civil defence science and thus reinforced.

Control

Mass panic is tied to a loss of control: the individual losing control over themselves and authorities losing control over populations. With maintaining government authority as a top priority, control of the population also quickly became a main priority for the Senior Committee, and it delegated the issue to the CDC. Val Peterson agreed and even stated in a Civil Defence meeting that control of the population was the most important aspect of *any* defence, civil or military. 'In his own country they were still very dubious regarding public reaction in the event of a war'.[38] The message was also employed at top level. Leaving office in 1957, NATO General Secretary Lord Ismay stated that it was the 'bounden duty of every Government to do its utmost' to ensure that in a war, 'the civil population and civilian activities would be efficiently controlled and directed'.[39]

Hence, control of the population was one of the first things the Senior Committee asked the CDC to review. Hodsoll responded with a September 1956 memorandum in which he earnestly stated that it was not something they had hitherto considered or even defined. In his definition, control meant having the population '(a) maintain their morale; (b) do what they are told, (c) do everything possible to help themselves'. The reasoning behind putting morale before obedience was that Hodsoll thought it was a prerequisite for exercising 'reasonable control' over people. Once people lost their spirit, it 'might easily result in the population degenerating into an unorganized and undisciplined mob, with all the consequences which that would imply'. While admittedly not previously having dealt with the issue, Hodsoll had quickly learned the language of this particular imaginary and its dystopia of the uncontrolled mob.

To sustain morale—and thereby control—Hodsoll, in line with the existing research consensus, considered it 'of cardinal importance' that the population be prepared and educated in peacetime. This was, incidentally, also all that NATO could do. Procedures for 'actual control'—execution of orders and regulations—were a matter for the individual countries' executive powers.[40] But as these functions may also be halted during a war, Hodsoll underlined that 'control, to be successful, must rely in the main on the self-discipline of the population, their ready acceptance of whatever orders are issued and their co-operation in such a way as to make such orders succeed'. Ideally, control was not to be enforced by already strained authorities but exercised from within subjects themselves. Self-help was 'a relief for the trained forces', but—also in line with existing research—he recommended some presence of authority to keep up morale *and* control. Martial law might be necessary in 'very adverse circumstances', but he recommended it only as 'a last resort'.[41]

Throughout 1957, several member countries submitted memoranda to the CDC on the issue. With slight variations, their focus was roughly the same: peacetime education and training and wartime establishment of authority and communication.[42] In an attempt to further structure these efforts, Hodsoll, in a January 1958 memorandum, divided measures into three situations: peacetime, a period of tension and war. Education and preparation should be done in peacetime, as surprise by the nature of nuclear war 'might rapidly [...] lead to panic and disorder'. If there were a period of pre-war tension, governments should issue preparatory instructions. In war, communications should be established as quickly as possible, especially to spread the message that 'the Government is still in control and knows what it is doing'. This should be backed by visible action and presence by local authorities or civil defence forces.[43] The sudden and overwhelming focus on control in the CDC made the Turkish representative remark at a 1957 meeting that some extent of population *protection* was presumably still needed—otherwise there would be no one left to control (Fig. 2.3).[44]

In 1958, NATO's Military Committee adopted a document known as MC70, which re-evaluated the alliance's force requirements and strategy. The document mentioned enemy psychological operations against morale and suggested the necessity of forces 'for maintaining order behind the combat zone', without going into further detail.[45] MC70 was an elaboration of MC14/2 of May 1957, which stated as a 'Defence Principle' under the North Atlantic Treaty, that

> [i]t is a responsibility of national authorities to develop plans and measures which will ensure the continuity of governmental control following a sudden outbreak of hostilities and will also ensure the maintenance of civilian morale coupled with the ability to prosecute the war to a successful conclusion.[46]

The North Atlantic Treaty itself, however, does not mention morale or continuity of government. One previous strategic document, MC48/1 of

Fig. 2.3 A vision of a government that is in control and knows what it is doing
In the 'Alert America' exhibition, people were assured that civil defence personnel would have a cool overview, guide citizens to proper locations and inform them what to do (https://www.digitalcommonwealth.org/search/commonwealth:h128rn65f)

December 1955 mentioned it in more lofty terms. Thus, the thermonuclear era had worked the issue of morale and control into official NATO strategy documents. This fact was taken as further confirmation by the CDC that an increased focus on 'prevention of panic' was warranted and that 'nations must be prepared to maintain order on their Home Fronts'.[47]

But once again, studying and making recommendations was all the CDC was able to do.[48] While the Senior Committee encouraged these studies it also recognised that the issue itself lay beyond the scope of 'purely civil defence [...], since it entered into the maintenance of public order and consequently was related to the maintenance of government control'.[49] As was once again established, actual maintenance of law and order was a job for the national governments.[50] The Senior Committee would, however, in the following years continue to urge nations to prioritise planning for maintenance of government control, including emergency laws and preparation.

Although not equipped with authority to execute it, the CDC continued studying and discussing related issues such as wartime command, control and communications, assistance of armed forces to civil defence, and the role of central and local government employees in war. As such, the element of (controlling) human reactions had made its way into emergency planning with the thermonuclear era.

Conclusion

When authorities imagined war, it wasn't always with the greatest of confidence in their populations. Authorities feared that citizens' fear would express itself in panic, which was considered dangerous and destructive. The need to tackle and control the base instincts of the population was older than, but intensified by, the advent of nuclear and thermonuclear weapons. In the early 1950s, fear of panic thrived especially with American civil defence planners, but the H bomb spread it to Europe. While it might not be possible to pinpoint an exact origin, we see a consistency of Western cultural thought dating back to the establishment of the modern psychological and social sciences. The civil defence dystopia of the uncontrollable mob was culturally consistent with earlier images, as was the belief (or hope) that this was somehow manageable by science. The nuclear and thermonuclear ages were thus negotiated into pre-existing frameworks of both civil defence and desired social order and were informed by research preceding it.

While the nuclear apocalyptic was (re-)negotiated into these lines of thought, the advent of the H bomb added a new sense of urgency. Prior to the establishment of the Senior Committee, the CDC and articles in the Civil Defence Bulletin had often enough underlined the importance of morale, but never entertained scenarios of chaos and human liability to the extent that it was done in the mid to late 1950s. Morale, having previously mainly been about spirit and support, was now directly tied to control. From not having been articulated at all, control of populations was now catapulted into top priority—especially as civil emergency planning emerged with its focus on maintaining government authority. In this particular nuclear dystopia, losing the lives of the population would be bad. Losing control over them would be even worse.

Imaginaries are mostly group achievements (Jasanoff 2015: 25). So was this one: it was formulated, shared and distributed by a civil defence community not only informed by, but equipped with, its own scientists, media and advisors. With nations employing their own scientists and experts, as most national civil defence directorates did, this also secured the continued co-production of this imaginary. In this case it also received the backing of the mighty NATO, which not only worked to extend the sociotechnical imaginary back to national governments, but also served as a forum for spreading and solidifying it.

But there were limitations. NATO's Civil Defence and Senior Civil Emergency Committees never faced up to a full panic scenario. While experts talked about physical measures like sedating or even fixating people, official documents consulted for this article only hinted at physical measures such as riot control or martial law. This was not solely due to unwillingness to discuss an unpleasant topic, but also due to the fact that execution of such powers was a national affair. NATO could negotiate, streamline and seek to extend the imaginary, but its final embedding was up to national governments. Some embraced it, pouring millions into both material and non-material preparations,[51] whereas most renegotiated it into a local setting. And while the imaginary might be collectively held by authorities, how it fared in civil society, and the resistance it met, is another, no less interesting, question (which is explored in several chapters in this volume).

Viewing the culture of Cold War preparedness promoted by civil defence organisations as a sociotechnical imaginary makes sense—although in this case, one must insist that science is not only technological or material. Science is also humanities, social, political and psychological, and those were the primary sciences at work here. This particular sociotechnical imaginary was also not formulated as a utopia, but a dystopia—nuclear war and, with it, a total collapse of society and the humans in it. Any desirable images in here were those of well-behaved, adaptable and docile citizens, a best-case scenario within the worst.

But this does not only tell us something about how war was imagined. Sociotechnical imaginaries are often a function of existing power/social order and aimed at preserving it (Jasanoff 2015: 21, 25–26). And what the fear of populations tells us about Cold War Western society might be the most tangible lesson. Civil defence, as other research has pointed out, was as much about reassuring people and securing their support for the security politics *du jour* as it was about, as this chapter has shown abundantly, upholding the control of existing orders. If the docile, calm and obedient behaviour desired in war was extended 'backwards' into peacetime, both governments and planners were better off. The Cold War was also total, and the home front needed mobilisation in peacetime as well. Civil defence did not just serve the purpose of assuring the public that they would be able to 'do something'—it was also about making them do the 'something' which they were told to do.

Thus, civil defence and preparedness were not only narratives of a future war, but also a narrative of the desirable citizen at large. The fixation on upholding government control as the most important prerequisite for a nation's survival and the almost nightmarish visions of panic, chaos and destruction say a lot about another dystopia, one in which Western populations would no longer support the Cold War ideological order. Making people feel like they would be okay, it was hoped, would make them less likely to revolt, and that was a central tenet of civil defence as well as society in general. As such, the sociotechnical imaginary of preparedness culture had the purpose of supporting and enforcing existing power relations.

The imaginary had a function not only in war, but also in peace. Education and preparation were one side of a coin that had social control as the other.

Notes

1. *Collier's Weekly*, 21 August 1953, 99–109. https://www.unz.com/print/Colliers-1953aug21-00099/ [accessed 16 October 2020].
2. AC/23-R/5: Record of Civil Defence Committee meeting 9–10 February 1953. NATO Archives 01—The North Atlantic Council, AC/23 Series, 1953, Records.
3. C-M(52)27: Civil organization in North Atlantic Treaty countries in times of war. 5 June 1952. NATO Archives, 01, C Series, 1952, Memoranda.
4. U.S. Strategic Bombing Survey: The Effects of the Atomic Bombings of Hiroshima and Nagasaki. 19 June 1946. Truman Library, https://www.trumanlibrary.gov/library/research-files/united-sta

tes-strategic-bombing-survey-effects-atomic-bombs-hiroshima-and [accessed 16 October 2020].
5. AC/23(CD)-D/44: Report summarizing all answers, 28 May 1953.
6. *Operation Doorstep*, Periscope Film. https://archive.org/details/28072OperationDoorstep [accessed 16 October 2020]. AC/23(CD)-R/3: Record of meeting. 11 November 1953.
7. AC/23(CD)D/108: Experimental atomic explosion in Nevada. 10 June 1955, Operation Cue (1955). https://www.youtube.com/watch?v=zU9lCKDzKSY [accessed 16 October 2020].
8. Wills (2019). On the role of civil defence films in constructing imaginaries, see the chapters by Hogg and Bennesved and Sylvest, this volume. On the normalisation of nuclear weapons, see also Hogg.
9. AC/23(CD)R/4. Record of meeting. 17 May 1954.
10. Brev fra Schultz til Dahl, 5 October 1954. Rigarkivet (Danish National Archives) 1261: Beredskabsstyrelsen, Civilforsvarsdirektør E. Schultz' embedsarkiv, 63: Breve 1950–61, læg 1.
11. 'Nogle hovedindtryk af amerikansk civilforsvar', 10 January 1955. Schultz, Materiale vedr. udlandet 1947–1983, 217: USA 1951–1960 mm., USA 1951–1955.
12. AC/23(CD)R/5: Record of meeting, 15–16 November 1954.
13. AC/23(CD)D/98: Civil Defence and Thermo-Nuclear Weapons, 25 November 1954; *Orientering* 3, 19 March 1956. Schultz, Mødereferater, diverse udvalg 1947–1983, 195.
14. C-M(55)48(Final): The application of the new assumptions to the work of the emergency planning committees. 27 May 1955.
15. C-M(55)48(Final): The application of the new assumptions to the work of the emergency planning committees. 27 May 1955.
16. AC/23(CD)D/112: Application of the new planning assumptions to the work of the Civil Defence Committee, 3 August 1955; C-M(55)75: US proposal for reorganization of civil emergency planning, 20 August 1955; C-R(55)37: Council meeting 14 September 1955; AC/23(CD)D/110: Revised draft report in current work in the light of the new assumptions. 19 September 1955; C-M(55)95: Draft resolution, 3 November 1955 and C-M(55)100: Final resolution, 10 November 1955; C-R(55)50: Record of council meeting. 14 November 1955. For earlier drafts and discussion of these, see AC 95 Series.

AC/23(CD)D/124: Application of the new planning assumptions to the work of the Civil Defence Committee. 19 December 1955; AC/23(CD)D/124(Final). Application of the new planning assumptions to the work of the Civil Defence Committee, 19 April 1956; AC/23(CD)R/7: Record of meeting. 2 December 1955.

17. 'Field Marshal Montgomery warns on "great danger of hysteria" after attack', *NATO Civil Defense Bulletin* August 1955. Schultz, NATO-CF-Bulletiner 1947–1983, 201: 1954–1959.
18. Synspunkter på psykiske reaktioner under krig, *Orientering* 8, 24 September 1956. Schultz, Mødereferater, diverse udvalg 1947–1983, 195. Title and quotes translated from Danish by the author.
19. 'Synspunkter på psykiske reaktioner under krig', *Orientering* 8, 24 September 1956. Schultz, Mødereferater, diverse udvalg 1947–1983, 195.
20. 'Civil morale under the threat of nuclear attack', *NATO Civil Defence Bulletin* March 1958. Schultz, NATO-CF-Bulletiner 1947–1983, 201: 1954–1959.
21. 'Psychological considerations in Civil Defence Planning', *NATO Civil Defence Bulletin* March 1959. Schultz, NATO-CF-Bulletiner 1947–1983, 201: 1954–1959.
22. 'Civil morale under the threat of nuclear attack', *NATO Civil Defence Bulletin* March 1958. Schultz, NATO-CF-Bulletiner 1947–1983, 201: 1954–1959.
23. 'Psychological considerations in Civil Defence Planning', *NATO Civil Defence Bulletin* March 1959. Schultz, NATO-CF-Bulletiner 1947–1983, 201: 1954–1959.
24. 'Synspunkter på psykiske reaktioner under krig', *Orientering* 8, 24 September 1956. Schultz, Mødereferater, diverse udvalg 1947–1983, 195.
25. 'Maintenance of morale among civilians during a nuclear emergency', *NATO Civil Defence Bulletin* October 1957. Schultz, NATO-CF-Bulletiner 1947–1983, 201: 1954–1959.
26. 'Synspunkter på psykiske reaktioner under krig', *Orientering* 8, 24 September 1956. Schultz, Mødereferater, diverse udvalg 1947–1983, 195.
27. 'Civil morale under the threat of nuclear attack', *NATO Civil Defence Bulletin* March 1958. Schultz, NATO-CF-Bulletiner 1947–1983, 201: 1954–1959.

28. 'Psychological considerations in Civil Defence Planning', *NATO Civil Defence Bulletin* March 1959. Schultz, NATO-CF-Bulletiner 1947–1983, 201: 1954–1959.
29. 'Synspunkter på psykiske reaktioner under krig', *Orientering* 8, 24 September 1956. Schultz, Mødereferater, diverse udvalg 1947–1983, 195.
30. 'Psychological considerations in Civil Defence Planning', op. cit. Underlinings in original text.
31. 'Civil morale under the threat of nuclear attack', *NATO Civil Defence Bulletin* March 1958. Schultz, NATO-CF-Bulletiner 1947–1983, 201: 1954–1959. This was reflected by increasing 'realism' in at least Danish and Swedish civil defence films from the late 1950s, see Bennesved and Sylvest, this volume.
32. 'Synspunkter på psykiske reaktioner under krig', *Orientering* 8, 24 September 1956. Schultz, Mødereferater, diverse udvalg 1947–1983, 195.
33. 'Maintenance of morale among civilians during a nuclear emergency', *NATO Civil Defence Bulletin* October 1957. Schultz, NATO-CF-Bulletiner 1947–1983, 201: 1954–1959.
34. 'Psychological considerations in Civil Defence Planning', *NATO Civil Defence Bulletin* March 1959. Schultz', NATO-CF-Bulletiner 1947–1983, 201: 1954–1959.
35. 'Synspunkter på psykiske reaktioner under krig', *Orientering* 8, 24 September 1956. Schultz, Mødereferater, diverse udvalg 1947–1983, 195.
36. 'Civil morale under the threat of nuclear attack', *NATO Civil Defence Bulletin* March 1958. Schultz, NATO-CF-Bulletiner 1947–1983, 201: 1954–1959.
37. 'Synspunkter på psykiske reaktioner under krig', *Orientering* 8, 24 September 1956. Schultz, Mødereferater, diverse udvalg 1947–1983, 195.
38. AC/23(CD)R/10: Record of meeting, 12–13 March 1957.
39. NATO. April 1952 – April 1957. Text of Lord Ismay's Report to the Ministerial Meeting of the North Atlantic Council in Bonn, May 1957. NA 04, Public Diplomacy Division, NATO Publications, NATO Non-Periodical, 1957.
40. AC/23(CD)D/151: Control of civilian population under attack. 6 September 1956.

41. AC/23(CD)D/151: Control of civilian population under attack. 6 September 1956.
42. 1957: AC/23(CD)D/192 (Netherlands), AC/23(CD)D/195 (UK), AC/23(CD)D/200 (Belgium), AC/23(CD)D/202 (Turkey) and AC/23(CD)D/204 (Luxembourg). 1958: AC/23(CD)D/244 (Germany).
43. AC/23(CD)D/216: Control of the Population, 21 January 1958.
44. AC/23(CD)R/11: Record of meeting, 19–20 November 1957.
45. MC70: The Minimum Essential Force Requirements, 1958–1963. 29 January 1958. NA 03, MC Series, 1958.
46. MC14/2 (Revised) (Final Decision). Overall Strategic Concept for the Defense of the North Atlantic Treaty Organization Area. 23 May 1957.
47. AC/23(CD)D/297: Civil Emergency Planning Assumptions—The Effect of the New Military Appreciation. 16 October 1958.
48. AC/23(CD)D/293: Suggestions by the United Kingdom delegation on heading under which the Civil Defence aspects of 'Control of the population' might be studied by NATO. 12 February 1959; AC/23(CD)D/193: Fallout control demonstration by the United Kingdom delegation. 2nd September 1957; AC/98-R/4: Record of meeting, 15–16 October, 30 October 1957.
49. AC/23(CD)N/40: Control of the population. October 1958. AC/23(CD)R/13: Record of meeting 25–26 November 1959.
50. AC/23(CD)D/283: Control of civilian population. 21 November 1958.
51. See especially Bjørnsson (2020), Frandsen (2020), Rimmen Noe et al. (2017), Aagard (2018) and the chapters by Farbøl and Bennesved and Sylvest, this volume.

References

Unpublished Sources

RA: Rigsarkivet (Danish National Archives).
1261: Beredskabsstyrelsen, Civilforsvarsdirektør E Schultz' embedsarkiv.
Breve 1950–1982, 63: Breve 1950–1961.
Mødereferater, diverse udvalg 1947–1983, 195.

Materiale vedr. udlandet 1947–1983, 217: USA 1951–1960 mm. USA 1951–1955.
NATO-CF-Bulletiner 1947–1983, 201: 1954–1959.

Online Sources

NA: NATO Archives—www.archives.nato.int
01: The North Atlantic Council
AC/23: Committee on Civil Organization in Time of War
AC/23(CD): Civil Protection Committee
AC/95: Working Group to Examine the United States Proposal for Reorganization of the Civil Emergency Planning Structure
AC/98: Senior Civil Emergency Planning Committee
C—North Atlantic Council
03: Military Committee
MC—Military Committee
04: International Staff / International Secretariat Public Diplomacy Division, NATO Publications

Colliers Weekly archive—https://www.unz.com/print/Colliers/

Truman Library
United States Strategic Bombing Survey: 'The Effects of the Atomic Bombs on Hiroshima and Nagasaki', June 30, 1946—https://www.trumanlibrary.gov/library/research-files/united-states-strategic-bombing-survey-effects-atomic-bombs-hiroshima-and?documentid=NA&pagenumber=1

Operation Doorstep—https://archive.org/details/28072OperationDoorstep
Operation Cue (1955)—https://www.youtube.com/watch?v=zU9lCKDzKSY

Literature

Aagaard Jensen, Sanne. 2018. *Nuclear-proof communications? The Cold War and the governance of telecommunications security in NATO and Denmark*. PhD Thesis, University of Copenhagen.

Biess, Frank. 2009. 'Everybody has a chance': Nuclear angst, civil defence, and the history of emotions in Postwar West Germany. *German History* 27: 215–243.

Bjørnsson, Iben. 2020. 'Stands tilløb til panik'. Civilforsvarspjecer som social kontrol. In *Atomangst og civilt beredskab*, eds. Marianne Rostgaard and Morten Pedersen, 65–102. Aalborg: Aalborg Universitetsforlag.

Blain, Daniel, Paul Hoch, and V. G. Ryan. 1945. A course in psychological first aid and prevention. *American Journal of Psychiatry* 101: 629–634.

Bourke, Joanna. 2006. *Fear: A cultural history*. London: Virago.

Cronqvist, Marie. 2008. Utrymning i folkhemmet: Kalla kriget, välfärden och den svenska civilförsvarskulturen 1961. *Svensk Historisk Tidskrift* 128: 451–476.

Cronqvist, Marie. 2012. Survival in the welfare cocoon: The culture of civil defence in Cold War Sweden. In *Cold War cultures: Perspectives on eastern and western European societies*, eds. Anette Vowinckel, Marcus M. Payk and Thomas Lindenberger, 191–210. New York: Berghahn Books.

Davis, Tracy C. 2007. *Stages of emergency: Cold War nuclear civil defense*. Durham: Duke University Press.

Drayer, Calvin S., Dale C. Cameron, Walter D. Woodward, and Albert J. Glass. 1954. Psychological first aid in community disaster. *Journal of the American Medical Association* 156: 36–41.

Frandsen, Bodil. 2020. Det centrale civile beredskab og REGAN Vest. In *Atomangst og civilt beredskab*, eds. Marianne Rostgaard and Morten Pedersen, 27–63. Aalborg: Aalborg Universitetsforlag.

Grant, Matthew. 2010. *After the bomb: Civil defence and nuclear war in Britain, 1945–68*. Houndmills: Palgrave Macmillan.

Grant, Matthew. 2016. The imaginative landscape of nuclear war in Britain 1945–65. In *Understanding the imaginary war: Culture, thought and nuclear conflict, 1945–90*, eds. Matthew Grant and Benjamin Ziemann, 92–115. Manchester: Manchester University Press.

Grant, Matthew, and Benjamin Ziemann, eds. 2016. *Understanding the imaginary war: Culture, thought and nuclear conflict, 1945–90*. Manchester: Manchester University Press.

Grossman, Andrew D. 2001. *Neither dead nor red: Civil defense and American political development during the early Cold War*. New York: Routledge.

Hennessy, Peter. 2010. *The secret state: Preparing for the worst 1945–2010*, second edn. London: Penguin Books.

Horn, Eva. 2016. The apocalyptic fiction: shaping the future in the Cold War. In *Understanding the imaginary war: Culture, thought and nuclear conflict, 1945–90*, eds. Matthew Grant and Benjamin Ziemann, 30–50.

Jasanoff, Sheila. 2004. The idiom of co-production. In *States of knowledge: The co-production of science and social order*, ed. Sheila Jasanoff, 1–13. London: Routledge.

Jasanoff, Sheila. 2015. Future imperfect: Science, technology and the imaginations of modernity. In *Dreamscapes of modernity*, eds. Sheila Jasanoff and Sang-Huyn Kim, 1–33. Chicago: Chicago University Press.

McEnaney, Laura. 2000. *Civil defense begins at home: Militarization meets everyday life in the fifties*. Princeton: Princeton University Press.

McMahon, Robert J. 2010. US national security from Eisenhower to Kennedy. In *The Cambridge history of the Cold War. Volume I: Origins*, eds. Melvin P. Leffler and Odd Arne Westad, 288–311. Cambridge: Cambridge University Press.

Nowak, Lars. 2016. Images of nuclear war in US government films from the Early Cold War. In *Understanding the imaginary war: Culture, thought and nuclear conflict, 1945–90*, eds. Matthew Grant and Benjamin Ziemann, 260–86. Manchester: Manchester University Press.

Oakes, Guy. 1994. *The imaginary war: Civil defence and American Cold War culture*. Oxford: Oxford University Press.

Rimmen Noe, Bjarke, Andreas Lie Stokbro, and Andreas Juel Høj. 2017. *Tryghed på tryk: Det danske presseberedskab under den kolde krig*. Master's Thesis, Aalborg University.

Rose, Kenneth D. 2001. *One nation underground: The fallout shelter in American culture*. New York: New York University Press.

Rostgaard, Marianne, and Morten Pedersen, eds. 2020. *Atomangst og civilt beredskab: Forestillinger om atomkrig i Danmark 1945–1975*. Aalborg: Aalborg Universitetsforlag.

Wills, John. 2019. Exploding the 1950s consumer dream: Mannequins and mushroom clouds at Doom town, Nevada test site. *Pacific Historical Review* 88: 410–438.

Ziauddin, Silvia Berger. 2017. (De)territorializing the home: The nuclear bomb shelter as a malleable site of passage. *Environment and Planning D: Society and Space* 35: 674–693.

Ziemann, Benjamin. 2016. German angst? Debating Cold War anxieties in West Germany, 1945–90. In *Understanding the imaginary war: Culture, thought and nuclear conflict, 1945–90*, eds. Matthew Grant and Benjamin Ziemann, 116–139. Manchester: Manchester University Press.

CHAPTER 3

The Imagined Disastrous: West German Civil Defence Between War Preparation and Emergency Management 1950–1990

Jochen Molitor

There is a ubiquitous quality to the term 'disaster'.[1] While it is certainly used to refer to worst-case scenarios or—adhering to a more technical definition—to events which exceed locally available resources (Münkler 2013: 135), it is probably more often utilised to describe mere inconveniences, especially in everyday life. The inherent character of the expression seems to encourage its use in vastly different circumstances ranging from the outrageous to the casual, from the horrific to the banal (Münkler 2013: 95). Without denouncing the pressing factuality of many catastrophic events, as some scholars tend to do,[2] an event does not necessarily have to be disastrous in and of itself to be labelled as such, and a disaster may just as much be considered a product of communication than of the forces of nature or the dangers of technology (Imhof 2004: 145). It is this *relativity* and *context-dependency* which has encouraged

J. Molitor (✉)
University of Cologne, Cologne, Germany

M. Cronqvist et al. (eds.), *Cold War Civil Defence in Western Europe*,
https://doi.org/10.1007/978-3-030-84281-9_3

scholars to be mindful of the term and observe the specific circumstances of its use. Beyond influencing two of the more recent general studies on contemporary West German[3] history, which has been characterised as a 'search for security' (Conze 2009) in a 'republic of fear' (Biess 2019), questioning *who* prepared *when* and for *what* 'disasters' seems highly instrumental for studying the civil defence endeavours of the Federal Republic of Germany. Accordingly, this chapter will focus on retracing the nature of the dystopian sociotechnical imaginaries (see the introduction to this volume) put forth by West German civil defence institutions, how they changed over time and what measures were taken to counter them. Using the term 'imagined disastrous' to stress both the act of imagining and that of labelling, it shall be argued that the relativity of the employed terminology allowed civil defence officials to model and adapt their proposed sociotechnical imaginary to befit the amount of preparation actually attainable by them.

Disaster as Rhetoric, War as Reality: Early West German Civil Defence 1950–1968

The Second World War and the experience of national defeat are frequently referred to as 'the disaster' (*die Katastrophe*) in early sources of the Federal Republic. Since labelling something as such is usually connected to the assumption that what has happened was the unintentional result of external forces (Münkler 2013: 97), this seems to reflect the widespread tendency of the new state to look for excuses rather than face accountability for the war (Moeller 2001: 94). During the 1950s, West German civil defence officials regularly employed similar terminology to obscure that the object of their preparations was, indeed, another military conflict, in which the divided Germany constituted the most prominent frontline between the opposing military alliances of the NATO and the Warsaw Pact. However, the underlying imagined disastrous was not so much based on the theoretical premises of nuclear warfare, but rather on first-hand experiences of the allied bombings of the recent past (Steneck 2005: 126–129).[4] Thus, early civil defence concepts had more in common with National Socialist air raid precautions than any kind of modern emergency management.[5] In addition, much of the same personnel was present (Diebel 2017: 36–37), and after talks started in 1948, official preparations intensified from 1950 onwards.

The foundation of a so-called Protective Commission (*Schutzkommission*) in 1951 aimed at funding scientific research on, for example, radiation sickness and the protective effect of different shelter types, while two organisations—the Federal Air Raid Protection Association (*Bundesluftschutzverband*, BLSV) and the Federal Agency for Technical Relief (*Bundesanstalt Technisches Hilfswerk*, THW)[6] aimed at educating volunteers in self-preservation measures and rescue operations. In late 1958, the basic organisation of a new West German civil defence system was finalised with the establishment of the Federal Office of Civil Defence (*Bundesamt für zivilen Bevölkerungsschutz*, BzB), which was to coordinate the various efforts.[7]

Although this organisational setup might in retrospect seem quite structured, its factual effectiveness beyond mere bureaucracy may well be questioned. A good example is the constant denial of policymakers to adhere to the central demand of basically all West German civil defence officials during the 1950s and early 1960s: for the widespread construction of shelters to protect the population at least from conventional bomb blasts and nuclear fallout. The first law including a prerequisite that all newly built houses were to have some sort of shelter facility was passed as early as October 1957. However, its main paragraph was immediately postponed and later dismissed due to its financial implications. Additional attempts to legally enforce or at least effectively support privately funded shelter construction suffered a similar fate; politicians of all parties simply did not want to commit to it (Steneck 2008: 82). Germany was thought to be a traumatised country with a populace bent on forgetting about the possibility of war rather than being constantly reminded of it (Molitor 2011: 44–50). Thus, even during the high point of construction efforts and across all crisis situations of the Cold War, West Germany could only provide public or private shelter spaces for approximately three per cent of its inhabitants (e.g. Geier 2003: 56). Civil defence officials glanced enviously at their neighbours in Switzerland, where basically everyone was provided for (cf. Berger Ziauddin 2017) and continued to insist: 'Without shelters, there can be no civil defence, and without them serving as the backbone, even the most effective organisational preparation will remain bogged in the merely hypothetical' (Sautier 1961: 47).

It was in this context that West German officials of the time often spoke of 'disasters' and shied away from the unpopular term 'war' as much as they could. However, in contrast to first aid or firefighting training, shelter construction eluded any serious kind of secondary use in times of peace and continuously brought home what the purpose of civil defence really was. Ultimately, the notion of any need for shelter construction revealed that the discourse around 'disasters' was merely a façade erected by officials who simply did not dare to tell the truth about their own profession. The separation of jurisdictions in the Federal Republic also highlighted this incongruity. While civil defence was, and still is, part of the national defence strategy, falling under the remit of the federal ministry of home affairs, disaster relief is considered a *local* endeavour, with the individual states governing their own varying regulations and institutions. Due to this legal separation, any talk of disasters by federal institutions such as the BzB was bound to come across as disingenuous, since they had no jurisdiction in these matters to begin with. Thus, the imagined disastrous of West German civil defence officials during the 1950s and early 1960s was clearly focused on a war against the forces of the Soviet Union and the Warsaw Pact. Initially it was imagined that this war would be fought with conventional weapons, but later it was envisaged as an all-out nuclear exchange for which West Germany, as the front state, would have to engage in total defence—reversing the National Socialist dictum of total war.

During the early 1960s, this line of thinking began to change. The widely debated paradigm shift in NATO's strategic doctrine from 'massive retaliation' to 'flexible response' (cf. Schmidt 1961 as a contemporary treatise from a German point of view), which promised 'appropriate' military reactions to perceived threats rather than total annihilation, made smaller conflicts more likely and meant that civil defence measures could actually prove somewhat effective (see also chapter by Farbøl this volume). Additionally, the idea that at least some such measures might also be helpful during natural or technical disasters became gradually more widespread. An early example was the North Sea flood of 1962, in which West German civil defence organisations—especially the THW, but also the BLSV—provided substantial assistance, despite not being officially responsible for local disaster relief (cf. Molitor 2014). While these organisations had traditionally spoken of disasters as a metaphor for a possible war, many contemporary sources show how much the actual disaster was now portrayed as warlike: the flood was described as an aggressor which had to be fought by daring heroes in a selfless struggle. This kind of work, for example within the uniform-wearing THW, might have offered many veterans of the former *Wehrmacht*—the army of National Socialist Germany—the opportunity to bask in a form of well-known martial rhetoric, now targeted at a non-human enemy and utilised for beneficial purposes, thus harking back to a cherished set of ideals purged of its former ideological content (Molitor 2014: 204–210).

Among other developments, events such as the flood of 1962 paved the way for the dual-use strategy put forth by civil defence officials from the mid-1960s onwards, resulting in probably the most significant West German civil defence law: the Act on the Expansion of Disaster Relief (*Gesetz zur Erweiterung des Katastrophenschutzes*, KatSG-68), passed in 1968. Misleadingly referring to civil defence as 'expanded disaster relief', the law was a political reaction to the dilemma that even a substandard civil defence system required cost-intensive investments, which—according to West German jurisdictions—could not be legally used except in the case of war, which everybody was desperately seeking to avoid. In contrast to that notion, the KatSG-68 demanded that federal authorities support the individual states' disaster relief efforts and, for example, grant them access to the THW as long as the states guaranteed the use of their equipment and institutions for civil defence efforts should the need arise. While this did not dissolve the traditionally split responsibility between federal civil defence and local disaster relief, it entangled them systematically in making one reliant on the other. The premise that preparedness in general had to be improved furthered the prioritisation of measures which could be useful in a variety of scenarios—alert systems, first aid education and so on—over prohibitively expensive and unwieldy genuine civil defence preparations such as shelter construction (cf. Diebel 2017: 76–79).

This trend towards what should later be called the 'all-hazards' approach can clearly be illustrated by comparing two of the key West German civil defence documents: the educational pamphlets *Jeder hat eine Chance* (*Everybody has a Chance*) and *Zivilschutzfibel* (*Guide to civil defence*). Both were published by the BzB and distributed for free to all households in 1962 and 1964, respectively. At the time, *Jeder hat eine Chance* was massively criticised for downplaying the nuclear threat, because it maintained that, depending on the distance of a nuclear detonation, even simple measures such as taking cover, protecting naked skin and looking away might prevent harm. The *Zivilschutzfibel*, however, fared somewhat better in the eyes of the public (Steneck 2005: 284–285; Biess 2009: 87–88). Released merely two years later, it contained all the relevant information of *Jeder hat eine Chance*, but while the original had clearly focused on the threat of war, the revised version (at fifty pages more than twice the size of its predecessor) began with the following statement, signed by the minster for home affairs at the time, Hermann Höcherl:

> Every day, we read or hear of unfortunate events. They threaten us at home or on the street. We're not even safe on our holidays. And, time and time again, humanity is threatened by natural disasters. [...] Even war, the biggest disaster of them all, is still a threat to many people of this world. (BzB 1964: 1)

War, as delineated in the *Zivilschutzfibel*, was not only labelled a disaster, but embedded within a larger framework of a multitude of possible dangers. Arguably, this played down the potential horror of an all-out nuclear conflict a lot more than *Jeder hat eine Chance*. Yet it also displayed a mindset which envisioned the disastrous as ranging from the trivial to the apocalyptic, with all possible dangers being somehow connected to each other, differing more in scope than fundamentally. Right in the middle of the pamphlet, a page headlined as 'An interim remark (It has nothing, absolutely nothing to do with war)' expressed this even more clearly:

> There will always be natural disasters. Floods – dyke breaches – earthquakes – massive fires. Don't you think that it would be hugely beneficial to know how to protect your life and your belongings? Think of the North Sea flood disaster for instance! Think of Hamburg! And what if a fire started in your own home? Maybe you could put your knowledge to good use for that! And if someone gets hurt at home or in a car accident, there'd be someone there to offer first aid – that would certainly be welcome! Even if you disregard all other arguments, these points should convince you to continue reading. (BzB 1964: 23, emphasis in original)

Again, the *Zivilschutzfibel* did not exclude the threat of war—it did in fact contain images of weaponry, of people taking cover and so on—but it also displayed a staggering array of possible alternatives, ranging from natural disasters to mere household fires, for which *the same kind of preparation* was declared useful. While its authors clearly tried to utilise disaster preparation as a less controversial means to get their point across, it must thus still be regarded as an early document publicly pushing the idea that civil defence should be *generally* useful.

As darkly comical as the brochure may seem from today's perspective, with the authors' insecurity about their potential readers' disposition unambiguously showing through, the *Zivilschutzfibel* did have a point, and the car crash photograph illustrating its 'Interim remark' was not only chosen to downplay nuclear annihilation. From the 1960s until the 1980s, more than 16,000 people died in West German traffic every year, reaching a tragic peak with over 21,000 casualties in 1970 (Kessel 2008: 52–53). If one compares these figures to the approximately 3,200 yearly traffic casualties during the 2010s, applying to a significantly more densely populated, reunified Germany, it seems obvious that not only civil defence was deficient during the 1950s and 1960s. General emergency management, too, failed to find answers to the 'modern' threats of the time, and with most people not wearing seatbelts, no first aid courses during drivers' training and no organised emergency medical system whatsoever, even smaller car accidents could easily end up being fatal. Keeping these observations in mind, the sometimes naïve-sounding arguments put forth by the *Zivilschutzfibel* should not be too easily dismissed. After all, it foreshadowed a significant change in both the civil defence authorities' sociotechnical imaginary and what kinds of activities would remain popular in West Germany from the late 1960s onwards: those which not only claimed to prepare for disasters, but actually *did* so in some way.

From the Sensationalist to the Pragmatic: Later West German Civil Defence

It has been stated that the move away from more dedicated measures to focusing on volunteer work and multipurpose skills rather than on building shelters effectively ended West German civil defence. Conversely, more recent studies suggest that it was this very move from the sensationalist to the pragmatic which saved it (Diebel 2017: 115–116). Because historians are, by and large, prone to telling good stories and have often shown considerable sympathy for contemporary peace movements which dismissed all kinds of civil defence as government propaganda at best and war mongering at worst (e.g. Garrison 2006; Klinke 2018), many accounts have focused on the more sensational aspects, glancing over its more pragmatic variants and all but overlooking its manifold connections to disaster relief. This has narrowed the viewpoint of many studies to the timeframe discussed thus far: the early Cold War period of the 1950s and 1960s. In the case of West Germany, however, it may be argued that its civil defence measures became (somewhat) successful *after* the big crisis situations of the early Cold War, with officials widely adopting the all-hazards approach from the late 1960s onwards.

In order to trace the later developments of civil defence in the Federal Republic, it is imperative to look beyond the state and its agencies. As already hinted at, civil defence in West Germany relied heavily on various relief organisations (*Hilfsorganisationen*).[8] Unlike the construction of shelters, the THW, for instance, was not abandoned in the 1960s, but thrived and flourished (Diebel 2017: 141–149). Today, it serves as one of Germany's foremost general relief organisations, providing refugee accommodation as well as giving aid during global pandemics.[9] The German Red Cross (DRK), too, was hardly ever criticised by even the most adamant peace activists, yet continuously supported civil defence and was quite successful in influencing its actual course during the 1980s (cf. Diebel 2017: 313–315). Just glancing at these two organisations—boasting current volunteer counts of c. 80,000 (THW) and 443,300 (DRK), respectively[10]—should clarify that there was (and is) a culture of preparedness that extends beyond lofty shelter construction plans put forth by a select number of officials but repeatedly cancelled before public discussion could even start.

This chapter now turns briefly to the development of West German disaster medicine (*Katastrophenmedizin*), not only to examine whether or not the imagined disastrous of physicians mirrored the general developments of West German civil defence, but also to further elaborate on the thesis that there was a move from a failed, expansive total defence approach to a partially implemented, more subdued one.

The Emergence of All-Hazards Disaster Medicine in West Germany 1956–1989

Physicians are interesting historical agents in civil defence research for several reasons. First of all, in the case of either a disaster or war, they are immediately involved at the frontline of the event and cannot enjoy the relative comfort of office professions. Most physicians do not work for the state but in hospitals or are self-employed, and they are generally well-respected, well-paid and used to a high degree of professional autonomy. However, since their work requires specialised training for which there is no easy substitute, all of them are required to help in times of crisis, especially those usually more occupied with treating the common cold than dealing with mass casualty situations. Depending on the perspective, physicians are thus both influential actors in a field that involves multiple professions combating for areas of exclusive jurisdiction (cf. Abbot 1988: 90) and a scarce and valuable resource for politicians or civil defence officials preparing for the worst.[11]

It is hardly surprising that the first physicians trying to establish some sort of medical civil defence in West Germany belonged to the military. In contrast to some of the civilian officials at the time, the imagined disastrous of most medical officers quickly focused on nuclear war.[12] Some certainly downplayed the potential destructiveness of this scenario and maintained that—if the planning was sound and the individual physicians masters of their craft—adequate preparations could be made (cf. Bayer 1962). However, the medical system (*Sanitätsdienst*) of the new West German army, the *Bundeswehr*, was deemed to be unfit to tackle such a task alone. Especially in the late 1950s and early 1960s, proposals were occasionally made to combine both the civilian and military spheres into an 'omnipotent' medical service, ready to face the most severe crisis situations and reflecting the sociotechnical imaginary of a shelter-based system of total defence. Still, even during the 1950s these proposals remained in the minority, since most military physicians deemed their powers to counteract the damage of the new weapons to be frightfully slim.

The 1962 large-scale NATO war exercise *Fallex* (*fall exercise*), renowned in German national history as the origin of the infamous *Spiegel* affair,[13] may serve as an example. During the exercise, military physicians professed that under the imagined circumstances of an all-out nuclear bombardment, they could maintain their service for a mere couple of days, if at all. Navy physician Karl-Wilhelm Wedel noted in his exercise diary that *Fallex* imagined 89 nuclear strikes on West Germany on day one, while official sources of the Ministry of the Interior mentioned 300 nuclear strikes within a six-day period.[14] Wedel warned that the general military's assessments of the casualties were far too low, stressed the large number of quick losses among the available medical personal, and pointed to an overall spread of panic and epidemics which would soon render any idea of organised medical help as completely unrealistic.[15] To allow for a wider field of possible decision-making, later NATO exercises, such as *Wintex* in 1975, usually referred to more restrained scenarios with only a limited use of nuclear weaponry. However, medical officers still held the opinion that they could only function with any degree of effectiveness as long as weapons of mass destructions were not used, and the criticism was made that even a single day of chemical warfare would render their services so 'heavily afflicted' that they would be unfit to offer further assistance.[16] While many military physicians presented themselves as 'uncomfortable reminders' within the general command structure,[17] it must be stressed that their largely pessimistic views did not stop them from preparing for the worst as well as they could. Befitting the prototypical soldier's duty not to give in easily, most of them remarked that *trying* to prepare was neither meaningless nor did it make a military conflict more likely, as was the opinion of various peace movements. On the contrary, a reasonable level of preparation, it was argued, could very well be for naught, but might still lead to some good. In one of his books on the subject, the *Bundeswehr*'s foremost radiologist Otfried Messerschmidt put it this way:

> Some might pose the question whether, in such a situation, anything should be done at all. Are not recommendations, as in this book for treating wounds, burns and radiation injuries under conditions of a massive holocaust, a pretense that a nuclear war can even be survived? Is it not naive to offer advice for a catastrophe that can be absolutely hopeless? If an impression of hopelessness results from reading this book, it is absolutely not the intention of the author. On the contrary, the information in this book is presented to spark initiative and imagination in a nuclear catastrophe in the firm belief that it will save human life. The will to help is not enough unless one knows what to do. [...] To be sure, nuclear war would be the most devastating event that could happen to mankind today. Should this catastrophe occur, against the will of all reasonable men, a medical service organization must stand ready to save as many lives as possible. (Messerschmidt 1979: xiv–iv)

The utilitarian notion to do whatever possible for as many as possible (cf. ZDv 49/50: 161) was a dominant motif of the early efforts of military physicians in these matters. Derived from traditional professional ethics, it served as a utopian counterpoint to the dystopian imaginary of nuclear annihilation. The medical method for achieving the necessary agency during the imagined disastrous was, however, highly controversial: triage—the process of selecting whom to treat and whom not to, in the case of injuries either too minor or too grievous, at times when there are too few physicians to be able to care for an overwhelming number of patients (cf. Koslowski 1969: 47). This last-resort measure of emergency medicine, barely contested in most countries, reminded many

Germans of the murderous selections which had occurred in Auschwitz and other National Socialist concentration camps. Nevertheless, for the military physicians dealing with civil defence issues, developing more effective methods of triage was considered vital. For them, triage was not linked to Nazism, but to their own professional past of practising surgery on the battlefields of the Second World War.[18] Most of the early medical officers of the *Bundeswehr* had been part of the *Wehrmacht* as well and ultimately regarded medical civil defence as a means of keeping something 'honourable', which had helped to save the life of many a soldier, from a tainted past.

In comparison with the Federal Republic's general military force, its dedicated medical service had been founded belatedly (1956 as opposed to 1955) and was deemed inferior to the army, air force and navy until—depending on one's viewpoint—1957 or even 1965.[19] The prestige of military physicians was often considered lacking,[20] and early medical officers sought out civil defence as a possibility of re-entering a working relationship with their civilian colleagues in a field which would affect *all* physicians in times of national crisis.[21] Apart from military officers such as Gustav Sondermann or Kurt Groeschel, civilian officials, too, supported such cooperative efforts, as they feared the military might otherwise veer away from the civilian sphere as it had done in the lead-up to and during the World Wars. Hans Neuffer, inaugural president of the newly founded *Bundesärztekammer* (German Medical Association),[22] expressly supported the military physicians' efforts to counteract 'disaster situations' while insisting that they had to remain medical doctors first and officers second.[23]

It is difficult to assess whether such endeavours were motivated primarily by an effort to keep the young democracy and its fragile institutions on track or simply to heighten their own professional power—or both. While still enjoying substantial societal prestige and solid incomes, it took German physicians some time to distance themselves from a past in which they had played a problematic part at best. They had been heavily involved in all top-level National Socialist organisations. As many as 45% had been members of the NSDAP, 26% of the SA and 9% of the SS (cf. Freimüller 2001: 14), and increasing revelations of nefarious medical crimes motivated many to keep their ranks closed. It was believed that strong ties between the diverse fields of the profession would quell internal turmoil which, in the end, was not in anyone's interest, as it endangered the strength of the profession as a whole. From this perspective, disaster medicine, which medical civil defence became increasingly known as from the late 1960s onwards, was a viable subject for bringing physicians from various fields together to work on. While it was originally derived from war surgery, it could—in the eyes of its proponents—only serve its purpose if *every* physician knew how to adhere to its central premise: the 'radical compulsion to think differently' (Rebentisch 1979: 99), in other words to turn from individual medicine to mass medicine in times of large-scale emergencies, using triage to save as many lives as possible. It may seem ironic then, that for all these efforts the actual implementation of disaster medicine was most effectively fuelled not by the military, nor even by the threat of war, but by something which had thus far eluded everyone's imagination: nuclear power plants.

The first major working group of the *Bundesärztekammer* associated with the potential dangers of the encroaching civilian use of nuclear power was founded in 1973 by renowned radiologist Emil Graul. Its results were published in 1975 (Scientific Council of the *Bundesärztekammer* 1975) and met with large-scale resentment at the 1976 German Medical Assembly (*Deutscher Ärztetag*)—the so-called 'parliament of the medical profession' in West Germany. Not only were its findings deemed too positive overall, with little regard paid to the problems stemming from radioactive waste materials, but the fact that several members of Graul's group had been on nuclear industry payrolls was also criticised.[24] Because high-ranking officials of the *Bundesärztekammer* were wary of political controversies within its ranks, a potential follow-up panel was to focus exclusively on the medical implications of nuclear power plants while avoiding easily 'politicised' issues such as radioactive waste. Furthermore, it had to refrain from inviting physicians with close ties to the industry.[25] Apart from the nuclear industry, however, most physicians specialising in radiation sickness belonged to the military. Thus, this latter demand opened up the chance to realise the hopes for cooperation expressed early on by military medical officers. More than twenty years after it had been originally proposed, an eclectic group of well-respected civilian and military physicians began creating a national scheme for disaster medicine, among them: the first German anaesthesiologist, Rudolf Frey; the civilian burn injury specialist Leo Koslowski; the already mentioned military physician Otfried Messerschmidt, renowned for having trained with radiology luminary Hanns Langendorff; and Ernst Rebentisch, retired chief inspector of the *Bundeswehr*'s medical service, the highest rank any medical officer could achieve.

Two working groups followed Emil Graul's original one, the first led by radiologist Hanno Poppe (cf. Scientific Council of the *Bundesärztekammer* 1981) and the second by military surgeon Rebentisch. Although a detailed reconstruction of the actual meetings cannot be provided here, the mere titles of the respective groups are telling. Graul's panel had been called 'Risks of Nuclear Power Plants'—indeed a title which did not necessarily restrict its focus to medical issues. In contrast to that, Poppe's working group 'Emergency Response of Physicians in Nuclear Power Disasters' concentrated strictly on the medical aspects of such disasters and imagined a single scenario: a reactor meltdown, deemed highly unlikely but theoretically possible. The final group, led by Ernst Rebentisch, began its work in 1981, resulted in two major publications in 1986 and 1988[26] and was simply called 'Deployment of Physicians during Disaster Events', its title being all but stripped of political overtones while generalising the nature of the imagined disastrous to effectively include almost anything.

Rebentisch himself was very much aware of the implications. For him, the all-hazards approach which was dominating civil defence officials' thinking at the time seemed all but mandatory. Indeed, he and his peers rejected the possibility that a physician could seriously prepare for the aftermath of a reactor meltdown without also preparing for at least a limited nuclear attack. Whereas the scale was certainly envisioned to be different, as was the asset of an intact medical service during times of peace, the required professional preparation, learning about triage, treating burns, radiation sickness and so on, seemed to remain completely unchanged. Therefore, Rebentisch could and deliberately did refer to disaster situations to avoid mentioning the highly unpopular term 'war' while still aiming for some basic wartime preparation of physicians.[27] However, having served in a variety of alternative scenarios himself—he had led the medical operation during the Palestinian terrorist attack on the Israeli team of the 1972 Summer Olympics in Munich, for example[28]—he did not regard talking of disasters solely as a rhetorical cloak, either. Rather, he tried to push the less controversial topic of disaster relief instead of civil defence because, as already stated, the professional consequences *of improving both* were identical in his opinion.

The debates surrounding a proposed *Gesundheitssicherstellungsgesetz* (GesSG) may serve as an example for this approach. The name of this bill was—whether chosen deliberately or not—highly complex, potentially misleading and difficult to translate. *Sicherstellungsgesetze*—laws 'to safeguard something'—are meant to guarantee the function of specific key services in times of national crises, in this case focusing on health. The GesSG belonged to a set of emergency laws (*Notstandsgesetze*) which were publicly embattled during the late 1960s. As an approved expert, Ernst Rebentisch was asked by the *Bundesärztekammer* to comment on the individual paragraphs of the draft. Rebentisch agreed with some of the key measures proposed therein, such as the registration of all medical workers (including those not currently employed) and the right to assess medical facilities for their potential use in mass casualty situations. Nevertheless, he and his peers remained highly critical of the bill. The knowledge of physicians in all matters of preparation concerning both disaster and war—which in Rebentisch's view had to be included from the planning stage onwards—tended to be ignored rather than sufficiently heeded. Throughout all his annotations, however, one wish was clearly paramount: to erase the term *Verteidigungsfall* ('in the case of national defence', a common euphemism for war) as thoroughly as possible and substitute it with 'disaster'.[29] Again, while Rebentisch was aware of the diplomatic advantage of this proposed change in terminology, the reasoning went beyond rhetoric. On multiple occasions, he insisted that the existing framework of medical disaster relief was even worse off than respective war preparations, which could at least rely on the *Bundeswehr*, and although the federal GesSG bill had no immediate jurisdiction on disaster relief issues as such, Rebentisch hoped that it would be interpreted by the individual states as an urgent call for action to improve their own legal situation.[30] Moreover, because of the dual-use nature of civil defence as delineated by the KatSG-68, any measures the federal state provided to improve it could potentially be used by the individual states for disaster relief as well.

Naturally, this all-hazards approach was, and still is, controversial in and of itself. As was hinted at earlier, it is a generalisation, if not a misconception, that civil defence measures may most easily be realised in times of crises. For the West German case at least, the very opposite seems to have been true. During times of détente, laws could be passed, structures established which would have met with widespread public and parliamentary resistance during the high points of the East–West conflict.

The years surrounding the Cuban Missile Crisis and the construction of the Berlin Wall in the early 1960s, as well as the NATO Double-Track Decision of the early 1980s led to multiple failed civil defence endeavours, since politicians feared public backlash. The outspoken peace movements of the early 1980s, for instance, were highly active in West Germany. Most dangerous to Rebentisch and other chamber experts were dissenting, well-respected physicians such as Ulrich Gottstein or Horst-Eberhard Richter who labelled specific preparations—especially training in triage—as a professional brand of warmongering in line with National Socialist' genocidal politics. Organised as the West German section of the International Physicians for the Prevention of Nuclear War (IPPNW) and praised by being awarded the Nobel Peace Prize in 1984, they achieved a sizeable amount of popularity and influence during the first half of the 1980s. They never represented a majority of physicians in West Germany, though, and ultimately failed to prevent a majority of the disaster medicine workgroup's goals (for a case study of the Swiss IPPNW, see chapter by Marti in this volume). It was not only the return of a politics of détente during the mid-1980s, however, which led to their defeat. Parallel to the widely discussed publication of Ulrich Beck's seminal work *Risikogesellschaft* (*Risk Society*, cf. Beck 1986), the reactor meltdown of the Chernobyl power plant on 26 April 1986 offered disaster medicine an opportunity to shine. Since scenarios above mere accidents, but below all-out nuclear war had always been those for which civil defence preparation might offer some kind of solace, physicians such as Rebentisch could now point to Chernobyl to support their call for additional preparedness while the West German section of the IPPNW gradually started to criticise nuclear power just as much as nuclear weapons.[31]

In 1989, the KatSG-68 was somewhat revised, and though its changes—brought about by the, again, awkwardly named *Katastrophenschutzergänzungsgesetz* (Amendment to the Disaster Relief Act)—could hardly be considered revolutionary, this revision realised some of the most important points professionals such as Ernst Rebentisch had been demanding for many years. Physicians were now integrated into crisis units from the planning stages onwards, and senior emergency physicians were given full authority over medical issues.[32] Over time, teaching triage and general disaster medicine became less controversial—though never unanimously supported—proving worthwhile in many a humanitarian relief operation, most of them abroad, during floods, earthquakes and—indeed—military conflicts.[33] However, contrary to the ambitions of

Rebentisch and his peers at the time, disaster medicine never emerged as a separate field of medical studies but became a part of emergency medicine, from which it does not fundamentally differ at the present time, apart from in its scope and severity.

Conclusion

For the year 1988, Martin Diebel states that the combined financial efforts at federal and state level for their dual-use 'expanded disaster relief' amounted to 3.7 billion Deutsche Mark—a sizeable sum indicating that civil defence had changed rather than failed in West Germany, as had the underlying sociotechnical imaginary of its proponents.

In the most prominent front state of the Cold War, split apart between opposing military alliances and hosting thousands of nuclear warheads, concepts of total defence had a hard time being accepted. Even the most expansive measures could not offer any guarantee, and ultimately, officials had to admit that everything hinged on the hope that the enemy simply might not *want* to eradicate West Germany. To be sure, many people chose to remain ignorant of the possible dangers out of comfort or denial, but some also remembered vividly what kind of fury could be unleashed by conventional bombardment alone. And while the equations between National Socialism and medical prioritisation processes regularly put forth by members of the West German IPPNW may seem absurd from today's perspective, if not downplaying the actual holocaust, they attest to the paramount role national history played in rendering expansive civil defence measures precarious to even utter. Additionally, every year that nuclear war did *not* happen raised the perceived necessity to focus on scenarios that did. The North Sea flood of 1962 was not hypothetical but a reality, and it did not stand alone. The Ramstein air show disaster of 1988, for instance, resulted in dozens of casualties and led to massive criticism of the poor state of disaster medicine and general emergency preparedness. Chernobyl motivated even IPPNW activists to publish a pamphlet reminiscent of the 1960s' civil defence brochures,[34] lending credence to the saying that during an actual crisis, even the worst critics of civil defence will try to seek shelter (Schneider 1962: 78).

Ultimately, civil defence in West Germany did not so much profit from the crises situations of the Cold War as it profited from peace. The respective officials and institutions who had initially talked of 'disasters' to avoid mentioning their real objective gradually started to believe their own rhetoric from the mid-1960s onwards. In a society growing ever more complex, the imagined disastrous expanded, including more and more scenarios of varying severity, all of them unlikely, but none impossible. Just as the discussions surrounding disaster medicine became more pragmatic and less steeped in moralising subtext, the underlying sociotechnical imaginary of general civil defence was gradually brought in line with attainable countermeasures that focused on volunteer relief work and medical training rather than digging holes in the ground.

Notes

1. This article focuses on German literature and sources. Although ‘catastrophe’ might seem to be the more literal translataion of the German term *Katastrophe*, its actual use is much more akin to ‘disaster’, as can be showcased by numerous word pairings such as *Katastrophenschutz* (‘disaster relief’), *Katastrophenfilm* (‘disaster movie’) and so on. Thus, ‘disaster’ was chosen as the most appropriate translation.
2. Both the writings of Agamben (cf. 2001: 8–14) and studies belonging to the theory of ‘securitisation’ (cf. Buzan et al. 1998 as an introduction) tend to support the notion that very few disastrous events require additional executive powers, but they are often abused by governments to this effect, e.g. by proclaiming states of emergency to bypass parliamentary control.
3. In matters of civil defence, the German Democratic Republic had more in common with the Soviet Union than with West Germany (cf. Müller 2009).
4. While the nuclear attacks on Hiroshima and Nagasaki provided a grim database from which to extrapolate the effects of larger scale conflicts, such deductions remained theoretical. Moreover, relevant research was not yet widely available in Germany during the 1950s.
5. See also the chapters by Bjørnsson, Hogg and Farbøl in this volume.
6. The BLSV, originally called *Luftschutzverein*, was founded as early as 1946, while the THW followed in 1950. They were recognised as federal institutions in 1951 (BLSV) and 1953 (THW).
7. For more detailed information on the genesis of the central institutions of German civil defence, cf. Steneck (2005) and, for the BzB, Geier (2003).

8. Many authors have underlined Germany's reliance on private associations (*Verbände*) in order to organise its politics, economy and culture (cf. Sebaldt and Straßner 2004). It may not even be widely known to the public, but although both disaster relief and general medical services appear to function as professional entities, they are, in fact, heavily reliant on the volunteer workers of various private associations (e.g. the German Red Cross, the *Arbeiter-Samariter-Bund*, the *Johanniter-Unfall-Hilfe*, the *Malteser-Hilfsdienst*, but also the THW and countless local fire departments). The consent of these associations proved to be a necessity for any civil defence reform. Their voluntary nature was certainly instrumental in achieving that status: politicians hardly dared to act against their will because they feared losing the only workforce they had—a workforce which, being unpaid, could easily quit at any moment.
9. Technisches Hilfswerk (ed.), Flüchtlingshilfe: Herausfordernde Aufgabe für das THW. Mission report, 18 September 2015, https://www.thw.de/SharedDocs/Meldungen/DE/Einsaetze/national/2015/09/meldung_006_ fluechtlingshilfe.html?param = 1,530,411,514,789&param3 = jsoff; Corona-Einsatz: THW-Kräfte unterstützen Gesundheitsämter. Press statement, 11 November 2020, https://www.thw.de/SharedDocs/Meldungen/DE/Einsaetze/national/2020/corona/meldung_060_corona_einsatz_1111.html;jsessionid=4FF40CE73B12796A33349214 5257BDEC.1_cid379?nn=924754 [both accessed 13 November 2020].
10. Cf. recent data taken from online press releases based on the 2019 yearbooks of both organisations: https://www.thw.de/SharedDocs/Meldungen/DE/Pressemitteilungen/national/2020/05/pressemitteilung_002_thw_jahresbericht_2019.html; https://www.drk.de/presse/pressemitteilungen/meldung/jahresbericht-2019-zahl-der-ehrenamtlichen-helfer-beim-drk-steigt-weiter/#:~:text=Das%20Deutsche%20Rote%20Kreuz%20%28DRK%29%20verzeichnet%20erneut%20einen,8.500%20auf%20insgesamt%20,443.300%20Helferinnen%20und%20Helfer%20gestiegen [accessed 12 November 2020].
11. For more detailed sociological remarks concerning physicians and other 'professionals', see the writings of Freidson (e.g. 1986).

12. IV H an Abteilungsleiter IV vom 13.5.1957, Betr.: Organisation des Führungsstabes des Gesundheits- und Sanitätswesens der Bundeswehr. BArch-MA, BW 24/20790.
13. The affair is usually considered a cornerstone in establishing the early Federal Republic's freedom of the press. It involved the political magazine *Der Spiegel* being accused of treason because it had published an article containing details about the German military's poor performance during Fallex 62 (cf. 'Bedingt abwehrbereit', *Der Spiegel* 41/1962, http://magazin.spiegel.de/EpubDelivery/spiegel/pdf/25673830).
14. Erfahrungsbericht über die NATO-Herbstübung 'FALLEX 62' vom 6.–27.9.1962; Anlage zu BMI VII B 5 (Gekürzte Fassung), p. 10. DRK, 4274.
15. Üb-Kriegstagebuch Nr. 1—KNDS-SAN FÜ. BArch-MA, BW 24/173.
16. WITEX SILVER/InSan II 1, Lagebeitrag Sanitäts- und Gesundheitswesen vom 13.3.1975. BArch-Ma, BW 24/5432.
17. Gustav Sondermann, 'Über das Leitbild des Arztsoldaten', in *Gespräch über militärärztliche Fragen vom 22. bis 24. Oktober* [1956] *an der Evangelischen Akademie Loccum*, p. 28. BArch-MA, BW 51/1.
18. C.f. Schreiben von Hermann Kater an Karsten Vilmar vom 28.8.1983. BArch, B 417/146.
19. The medical service became an independent part of the German military in 1957, but its commander (*Inspekteur*) was deemed inferior compared to the others until 1965.
20. c.f. Georg Meyer, Aktennotiz vom 7.3.1979. BArch-MA, BW 24/7796.
21. Kurt Groeschel, Gedanken zur Koordination von aktiven San. Offizieren und Ziviärzteschaft, 20 June 1956. BArch, B 389/112.
22. The official translation—German Medical Association—is somewhat misleading, as the *Bundesärztekammer* is not a typical association based on voluntary membership, but a 'chamber' (*Kammer*). All German physicians are organised within a so-called chamber system granting them professional autonomy as well as some federally guaranteed privileges. Membership in one's respective state-level chamber (*Landesärztekammer*) is mandatory for every practising physician. The delegates of these state-level chambers constitute the *Bundesärztekammer*, which serves as both a lobby

organisation and a professional think-tank usually focusing on issues which demand federal standards, such as medical ethics.

23. Hans Neuffer, 'Die deutsche Ärzteschaft zum Aufbau des Sanitätswesens der Bundeswehr', in *Gespräch über militärärztliche Fragen vom 22. bis 24. Oktober* [1956] *an der Evangelischen Akademie Loccum*, p. 18–23. BArch-MA, BW 51/1.
24. Cf. Das Risiko der Kernkraftwerke, in *Deutsches Ärzteblatt* 1/1973, p. 21–23.
25. Hanno Poppe, 18. Plenarsitzung des Wissenschaftlichen Beirats der BÄK (5.5.1979, Köln, Bundesärztekammer), TOP 3 C–Arbeitskreis 'Ärztliche Hilfe bei Katastrophen', p. 1. BArch, B 417/864.
26. The first publications were the *Denkschrift Katastrophenmedizin* (*Position Paper on Disaster Medicine*, Bundesärztekammer 1986), a short pamphlet primarily designed to put pressure on the legislative process, and the second *opus magnum* by Rebentisch, the *Handbuch der medizinischen Katastrophenhilfe* (*Handbook of Medical Disaster Relief*), containing detailed instructions for a multitude of medical fields in varying scenarios (Rebentisch 1988).
27. Letter from Ernst Rebentisch to Heinz-Peter Brauer vom 8.2.1982. BArch, B 417/154.
28. Cf. Laudatio zur Verleihung der Paracelsus-Medaille an Prof. Dr. med. Ernst Rebentisch, https://www.bundesaerztekammer.de/fileadmin/user_upload/downloads/Lau-rebentisch.pdf [accessed 13 November 2020].
29. Letter from Ernst Rebentisch to Heinz-Peter Brauer vom 13.10.1980. BArch, B 417/160.
30. Cf. Ernst Rebentisch, Beantwortung des Fragenkatalogs zur Anhörung am 6.10.1989. BArch, B 417/2924.
31. Since this paper focuses on civil defence rather than its critics, it cannot offer a detailed analysis of the IPPNW. For a study on the German section cf. Kemper (2016). As with general peace movements, opinions of the IPPNW both then and now were and are mixed. On the one hand, they were credited with softening the dividing line between friend and foe during the Cold War and for offering educational information on the 'untreatable' horror of nuclear war. On the other hand, they were criticised for one-sidedly condemning the defence preparations of liberal democracies while turning a blind eye to the totalitarian nature of the USSR, which cruelly persecuted dissenting activists at home.

32. Cf. Entwurf eines Gesetzes zur Ergänzung des Katstrophenschutzgesetzes und anderer Vorschriften (Katastrophenschutzergänzungsgesetz—KatSErgG), Stand: 8.9.1988. FZH, 16–3 A/2.3.–7.
33. A former student of Leo Koslowski, surgeon Bernd Domres alone led approximately 50 missions abroad (e.g. in Cambodia, Haiti, Nigeria and Pakistan) and co-founded the German Institute for Disaster Medicine, cf. https://www.stiftung-difkm.de/team/kurzvitae/ [accessed 13 November 2020].
34. Lutz Castell and Katharina Strecker (Institut für Umweltuntersuchungen Starnberg e.V.), Maßnahmen beim nächsten Super-GAU, 10 April 1987. FZH, 16–3 A/17.3–5.

References

Unpublished Sources

Federal Archives of Germany, in Koblenz (BArch).
- Holdings of the *Bundesärztekammer* (German Medical Association, B 417).
- Holdings of the *Hartmannbund* (Association of German Physicians, B 389).

Federal Archives of Germany—Military Archive, in Freiburg (BArch-MA).
- Holdings of the *Inspektion des Sanitätswesens der Bundeswehr* (Inspectorate of the Military Medical and Health Services, BW–24).
- Holdings of the *Deutsche Gesellschaft für Wehrmedizin und Wehrpharmazie e.V.* (German Society for Military Medicine and Military Pharmacy, BW 51).

Archive of the Research Centre for Contemporary History in Hamburg (*Forschungsstelle für Zeitgeschichte in Hamburg*, FZH).
- Holdings of the *Deutsche Sektion der IPPNW* (West German section of the International Physicians for the Prevention of Nuclear War, 16–3 A).

Archives of the General Secretariat of the German Red Cross, in Berlin (DRK).

Literatures

Abbot, Andrew. 1988. *The system of professions: An essay on the division of expert labor*. Chicago: Chicago University Press.

Agamben, Georgio. 2001. *Ausnahmezustand (homo sacer 2.1)*. Translated into German by Ulrich Müller-Schott. Frankfurt/M.: Suhrkamp.

Bayer, Alfred. 1962. Gedanken über die Sanitätstaktik in einem Verteidigungsfall. *Wehrmedizinische Mitteilungen*: 161–164 (part I), 178–180 (part II), 196–198 (part III).

Beck, Ulrich. 1986. *Risikogesellschaft*. Frankfurt/M.: Suhrkamp.
Berger Ziauddin, Silvia. 2017. Superpower Underground: Switzerland's rise to global bunker expertise in the Atomic Age. *Technology and Culture* 58: 921–954.
Biess, Frank. 2009. Jeder hat eine Chance: Die Zivilschutzkampagnen der 1960er Jahre und die Angstgeschichte der Bundesrepublik. In *Angst im Kalten Krieg*, eds Bernd Greiner, Christian Th. Müller and Dirk Walter, 61–93. Hamburg: Hamburger Edition.
Biess, Frank. 2019. *Republik der Angst: Eine andere Geschichte der Bundesrepublik*. Hamburg: Rowohlt (Published in English in 2020 as: *German Angst: Fear and democracy in the Federal Republic of Germany*. Oxford: Oxford University Press).
Bundesärztekammer, ed. 1986. Denkschrift Katastrophenmedizin. *Deutsches Ärzteblatt* 16: 1097–1111.
Bundesamt für zivilen Bevölkerungsschutz (BzB), ed. 1961. *Jeder hat eine Chance*. Bad Godesberg: BzB.
Bundesamt für zivilen Bevölkerungsschutz (BzB), ed. 1964. *Zivilschutzfibel: Informationen, Hinweise, Ratschläge*. Bad Godesberg: BzB.
Buzan, Barry, Ole Wæver and Jaap de Wilde. 1998. *Security: A new framework for analysis*. Boulder: Lynne Rienner Publishers.
Conze, Eckart. 2009. *Die Suche nach Sicherheit: Eine Geschichte der Bundesrepublik Deutschland von 1949 bis in die Gegenwart*. München: Siedler.
Davis, Tracy. 2007. *Stages of emergency: Cold War nuclear civil defence*. Durham: Duke University Press.
Diebel, Martin. 2017. *Atomkrieg und andere Katastrophen: Zivil- und Katastrophenschutz in der Bundesrepublik und Großbritannien nach 1945*. Paderborn: Ferdinand Schöningh.
Freidson, Eliot. 1986. *Professional powers: A study of the institutionalization of formal knowledge*. Chicago: Chicago University Press.
Freimüller, Tobias. 2001. Mediziner: Operation Volkskörper. In *Karrieren im Zwielicht: Hitlers Eliten nach 1945*, ed. N. Frei, 13–69. Frankfurt/M.: Campus.
Garrison, Dee. 2006. *Bracing for Armageddon: Why civil defence never worked*. Oxford: Oxford University Press.
Geier, Wolfram. 2003. *Zwischen Kriegsszenarien und friedenszeitlicher Katastrophenabwehr: Zur Entwicklung der zivilen Verteidigung in der Bundesrepublik Deutschland unter besonderer Berücksichtigung des Zivilschutzes und seiner Reformen vor und nach Beendigung des Kalten Krieges*. Doctoral dissertation, Marburg.
Imhof, Kurt. 2004. Katastrophenkommunikation in der Moderne. In *Katastrophen und ihre Bewältigung: Perspektiven und Positionen*, eds Peter Rusterholz and Rupert Moser, 145–163. Bern: Haupt.

Kemper, Claudia. 2016. *Medizin gegen den Kalten Krieg: Ärzte in der anti-atomaren Friedensbewegung der 1980er Jahre*. Göttingen: Wallstein Verlag.
Kessel, Nils. 2008. *Geschichte des Rettungsdienstes 1945–1990: Vom 'Volk von Lebensrettern' zum Berufsbild 'Rettungsassistent/in'*. Frankfurt/M.: Peter Lang.
Klinke, Ian. 2018. *Cryptic concrete: A subterranean journey into Cold War Germany*. Hoboken: Wiley.
Koslowski, Leo. 1969. Die Sichtung (Triage) der Verletzten bei Massenkatastrophen. *Katastrophenmedizin* (supplement to *Wehrmedizin*), 7th annual volume: 46–49.
McEnaney, Laura. 2000. *Civil defense begins at home: Militarization meets everyday life in the fifties*. Princeton: Princeton University Press.
Messerschmidt, Otfried. 1979. *Medical procedures in a nuclear disaster: Pathogenesis and therapy for nuclear-weapons injuries*. München: Thiemig.
Moeller, Robert. 2001. Remembering the war in a nation of victims. In *The miracle years: A cultural history of West Germany 1949–1968*, ed. Hannah Schissler, 83–109. Princeton: Princeton University Press.
Molitor, Jochen. 2011. *Mit der Bombe überleben: Die Zivilschutzliteratur der Bundesrepublik 1969–1964*. Marburg: Tectum.
Molitor, Jochen. 2014. Lehren für den Verteidigungsfall: Die Sturmflutkatastrophe von 1962 und der bundesdeutsche Zivilschutz. In *Die Hamburger Sturmflut von 1962: Risikobewusstsein und Katastrophenschutz aus zeit-, technik- und umweltgeschichtlicher Perspektive*, eds M. Heßler and C. Kehrt, 195–221. Göttingen: Vandenhoeck & Ruprecht.
Müller, Christian Th. 2009. Im Banne der Bombe: Überlegungen zu Luftschutz und Zivilverteidigung in der DDR. In *Angst im Kalten Krieg*, eds Bernd Greiner, Christian Th. Müller and Dirk Walter, 94–122. Hamburg: Hamburger Edition.
Münkler, Herfried. 2013. Revolution, Krieg & Katastrophe: Ein Diskurs über Domestizierung und Enthegung. In *Aufbruch ins Unversicherbare: Zum Katastrophendiskurs der Gegenwart*, eds Leon Hempel, Marie Bartels and Thomas Markwart, 95–139. Bielefeld: Transcript.
Rebentisch, Ernst. 1979. Katastrophenmedizin. *Wehrmedizinische Monatsschrift* 4: 97–100.
Rebentisch, Ernst. 1988. *Handbuch der medizinischen Katastrophenhilfe*. München-Gräfelfing: Werk-Verlag Banaschewski.
Rose, Kenneth. 2001. *One nation underground: The fallout shelter in American culture*. New York: New York University Press.
Sautier, Hermann. 1961. Der Weg des BLSV. *Zivilschutz* 25: 44–48.
Schmidt, Helmut. 1962. *Verteidigung oder Vergeltung: Ein deutscher Beitrag zum strategischen Problem der NATO*. Stuttgart-Degerloch: Seewald.
Schneider, Wolf. 1962. *Hat jeder eine Chance?* Neuenbürg: Osang.

Scientific Council of the *Bundesärztekammer*, ed. 1975. Stellungnahme zum Thema 'Gefährdung durch Kernkraftwerke'. *Deutsches Ärzteblatt* 41: 2821–2824.

Scientific Council of the Bundesärztekammer, ed. 1981. *Zur Frage der ärztlichen Versorgung der Bevölkerung bei Kernkraftwerksunfällen.* Köln: Deutscher Ärzte-Verlag.

Sebaldt, Martin, and Alexander Straßner. 2004. *Verbände in der Bundesrepublik Deutschland: Eine Einführung.* Wiesbaden: Verlag für Sozialwissenschaften.

Steneck, Nicholas. 2005. *Everybody has a chance: Civil defence and the creation of Cold War West German identity, 1950–1968.* Doctoral dissertation. Columbus, Ohio.

Steneck, Nicholas. 2008. Eine verschüttete Nation? Zivilschutzbunker in der Bundesrepublik Deutschland 1950–1965. In *Bunker: Kriegsort, Zuflucht, Erinnerungsraum*, eds Inge Marszolek and Marc Buggeln, 75–88. Frankfurt/M.: Campus.

Wolfrum, Edgar. 2006. *Die geglückte Demokratie: Geschichte der Bundesrepublik Deutschland von ihren Anfängen bis zur Gegenwart.* Stuttgart: Klett-Cotta.

ZDv 49/50: Die dringliche Kriegschirurgie. Übersetzung des NATO-Handbook *Emergency War Surgery* (1958). Bonn: Bundesministerium der Verteidigung. November 1961.

CHAPTER 4

Normalising Nuclear War: Narrative Scenarios, Imaginative Geographies and Sites of Leisure in 1950s Britain

Jonathan Hogg

'A roar of engines in the sky shattered the Sunday morning quiet, a squadron of jet planes swooped low over Morpeth Dock–and the largest fire service and civil defence exercise to be held in the North-West [UK] since the end of the war had started'. This newspaper report, published in *Birkenhead News* on 9 April 1952, went on to say that the civil defence 'exercise was a realistic test, and an imaginative attempt was made to stimulate the conditions under which the different services would have to operate and co-ordinate in war-time'.[1] In a briefing before the exercise, it was explained that 'as with all exercises, imagination must be brought into full play', which was an attempt to encourage participants to throw themselves into the exercise wholeheartedly.[2] The public description of this drill reflected 'the imaginative and rhetorical force of military spectacle'

J. Hogg (✉)
School of Histories, Languages & Cultures, The University of Liverpool, Liverpool, England
e-mail: jgh@liv.ac.uk

M. Cronqvist et al. (eds.), *Cold War Civil Defence in Western Europe*,
https://doi.org/10.1007/978-3-030-84281-9_4

(Rech 2015: 536), while also stressing the importance of both 'realism' and 'imagination.'

This representation of preparedness would become normalised in the early Cold War era. British citizens, via civil defence propaganda materials and official publications disseminated in the local and national press, were invited to imagine themselves as part of a new struggle which demanded a form of active psychological and physical engagement with the nuclear threat, and the acceptance of sustained militarisation in peacetime. As the jet planes roared over the River Mersey in North-West England as part of the Morpeth exercise, they acted out the imagined aerial attacks that would feature at the beginning of most civil defence 'narrative' scenarios in the 1950s. These scenarios acted as the starting point for training exercises carried out by civic leaders and volunteer personnel who also developed sites of leisure and forms of civic engagement that contributed to a culture of civil defence activism in the 1950s. The scenarios, and the physical responses to them in the exercises that followed, were also shaped by a shared knowledge of the spatial and geographical specificity of the region in which they were located.

In this chapter, civil defence is interpreted as one significant part of a process of mobilisation after the large-scale national development and institutionalisation of military nuclear technologies that occurred in Britain after 1945. One central aim behind British nuclear mobilisation was the creation of nuclear weapons and, following Jasanoff and Kim, once this initially secret technological development became public knowledge, a sociotechnical imaginary emerged which (1) downplayed the unique danger of nuclear weapons, (2) attempted to contain fear and encourage survivability and control, and (3) promoted deterrence ideology and the development of military nuclear technology as national virtues.[3] This imaginary pointed the way towards a desirable future, and was encouraged systematically by official means. Civil defence played a key role in 'embedding' this sociotechnical imaginary in British society, and I argue that the process of embedding this imaginary can only be properly understood with reference to localised social, geographical and discursive contexts.[4] Furthermore, I argue that the social impact and persistence of *nationwide* nuclear sociotechnical imaginaries cannot be fully understood without reference to the localised contexts in which civil defence was located and enacted.

In order to explore this further, a particular reading of Jasanoff and Kim's definition of sociotechnical imaginaries is applied to materials in

Wirral Archives (Merseyside, UK) and other relevant primary source materials from the period. These archival materials offer a glimpse into civil defence strategies in the North-West region of England, and especially the localised activities of volunteer activists and civic leaders on the Wirral peninsula. Wirral Archives hold a wealth of national and regional civil defence circulars, pamphlets and planning documents and, perhaps most usefully, regular correspondence between the Chief Clerk of Birkenhead Council, Mr. D. P. Heath, and the Home Office in London. We are afforded a glimpse of how national policies were enacted locally, and how the nuclear sociotechnical imaginary was replicated and made durable.

To trace the ways in which these localised contexts appear to have intertwined with and been shaped by a wider (officially maintained) sociotechnical imaginary, I do three things in this chapter. After brief discussions of general context and historiography, I explore some of the bespoke narrative scenarios that were created to frame civil defence exercises and analyse their public representation. Secondly, I focus on the sites of leisure and forms of civic engagement linked to civil defence activity more broadly. Third, I look to the concept of *imaginative geographies* (Said 1978; Gregory 1995) to further unearth the localised nature of sociotechnical imaginaries in this era.

Civil Defence in the UK

In the years following the conclusion of the Second World War, many British citizens faced severe social and economic hardship. Food rationing continued until 1954, and many urban communities struggled to deal with the destruction caused by the indiscriminate aerial bombardment that had become normal practice in the age of total war. A further shadow was cast over British life with the advent of the atomic bomb, and by 1949 the Soviet Union had tested their first weapon. Although the civil defence exercise at Morpeth Dock, Birkenhead—located on the Wirral peninsula in the North-West of England—was based on a non-nuclear scenario, it had become normal to devote time and resources to planning and performing realistic civil defence exercises in the early Cold War era. Indeed, nuclear and non-nuclear exercises were planned and run along similar lines: active nuclear preparedness dovetailed with older civil defence methods.[5] The news that the Soviets had successfully tested a weapon led to increased funding to civil defence, enabled by the Civil Defence Act of 1948, which by 1949 'empowered local authorities to appoint Civil Defence Committees and recruit members for the new Civil Defence Corps' (Campbell 1983: 71).

In this early atomic era—before the thermonuclear revolution, and before the Hall and Strath Reports in Britain shaped more realistic thinking about the likely impact of nuclear strikes on the British Isles (Hughes 2003)—atomic civil defence was rudimentary, and largely followed Second World War habits of thinking about survival and morale (Grant 2010: 44–51). This large-scale peacetime effort depended on volunteer recruits, and a significant Cold War propaganda campaign influenced British culture in the 1950s and 1960s. Citizens were persuaded of the necessity of Cold War participation, duty and responsibility both by the dissemination of government publications to civic leadership and then by advertisements placed in the national press, poster campaigns and recruitment films (Grant 2010; Hogg 2016).

This propaganda and recruitment drive was central to how the three main characteristics of the nuclear sociotechnical imaginary were embedded in British culture. This broadly discursive context—where particular ideas about nuclear weapons were disseminated by the government—was intertwined with the physical organisation of significant numbers of civil defence recruits in the 1950s.[6] By the late 1950s, civil defence organisation was conducted at sub-regional, regional and national level, and throughout the decade, a series of large-scale exercises were introduced as national policy, which would then be interpreted and organised by regional leadership.

Historiography: Civil Defence and Sociotechnical Imaginaries

In order to place civil defence activity within its localised social, geographical and discursive contexts, this chapter builds on a number of historiographical strands. Surveys of postwar Britain are plentiful, but approaches highlighting the powerful role of cultural memories of the Second World War (Noakes and Pattinson 2013) and the politics of reconstruction in the 1950s (Zweiniger-Bargielowska 2000) have especially influenced this chapter. Some potent and shared 'nation-building' imaginaries with longer traditions in British life that are also relevant to the history of civil defence include patriotism, nostalgia, militaristic masculinities and war commemoration (Anderson 1983). Temporally speaking, we might observe that these imaginaries stretch back in time, seeking solace in tradition or an imagined past. This is quite distinct from the future-oriented imaginaries of civil defence policymakers (and their representations), detailed in the work of David Monteyne (2011) and Joseph Masco (2008).

Histories of civil defence in the UK have focused on both the strategic 'façade' that the organisation represented (Campbell 1983; Grant, 2010) and localised histories of how Cold War policy impacted everyday life (Barnett 2015; Hogg and Brown 2019). Interdisciplinary research on atomic urbanism (Monteyne 2011) and imaginative geographies (Said 1978; Gregory 1995) suggest the importance of 'influential presentations of the world and its contours that are made possible by particular forms of knowledge about that world' (Farish 2010: xii). What many of these approaches to the Cold War have in common is an interest in the power dynamics at play in Cold War societies where, for example, 'nuclear spaces are constructed socially and politically' (Alexis-Martin and Davies 2017: 4). Jessica Douthwaite's recent oral history of civil defence personnel suggests that Cold War policy and ideology created permanent, durable qualities in government and society that were hard to resist. She argues that anti-nuclear protest faced an increasingly challenging task against this durable system, stating 'oral testimonies provide an invaluable means to demonstrate that where in its early years it made "sense" to critique nuclear security, anti-nuclear politics became increasingly irrelevant in an established, legitimised Cold War system' (Douthwaite 2018: 231). Furthermore, her thesis demonstrates the processes 'by which postwar Britain accepted and acclimatised to Cold War nuclear strategies' (Douthwaite 2018: 176).

Work that has examined aspects of the 'cultural Cold War' (Shaw 2005) and the role of the mass media in the British Cold War (Hill 2018) sit alongside work on the importance of assessing how, why and where nuclear anxieties were articulated in the 1950s (Hogg 2012). In British life, a form of 'nuclearity' existed that can be defined as 'a shifting set of assumptions held by individual citizens on the dangers of nuclear technology, assumptions that were rooted firmly in context and which circulated in, and were shaped by, national discourse' (Hogg 2012: 537). This can suggest that implicit understanding of the negative aspects of nuclear culture had become normalised throughout the 1950s and, further, official attempts were made to shape attitudes towards nuclear weapons (Hogg 2016).

This process, which we can identify through its discursive traces, can help us to make sense of how nuclear sociotechnical imaginaries became a normalised and persistent feature of Cold War life. Acknowledging a recent surge in interest in the analytical utility of 'sociotechnical imaginaries', the concept is conceived in this chapter as a powerful 'imaginative infrastructure' that serves 'the possibility of shaping terrains of choices and thereby of actions' (Sismondo 2020). These social constructs might have the potential to limit or encourage discussion in favour of an imaginary which serves a powerful agenda (Polleri 2020). As Robinson suggests, 'the collective shaping of imaginaries is not simply about creating communities, because they also allow states to more effectively control society and "other" those who do not share their same vision' (Robinson 2020: 3). As other contributors to this book note, imaginaries have the potential to 'shape meaning and behaviour', and it is important to analyse the cultural components that enable this.[7] Perhaps most closely aligned to the approach in this chapter is a recent attempt to trace the 'intricate relationship between technology and social change in relation to cultural memory and cultural perceptions' (Kalmbach et al. 2020), especially also considering their contention that it is often fear that provides the impetus for technological development.

The definition of sociotechnical imaginaries by Jasanoff and Kim cited in the introduction to this book is a description of a tiered and interlocking process of structure and agency. Taking inspiration from this, civil defence activities will be explored in this chapter as one crucial part of a useful 'collectively held, institutionally stabilised and publicly performed vision of a desirable future' that served to distort the reality of nuclear weapons as destructive and radioactive weapons of war. Crucially, this sociotechnical imaginary, animated as it was 'by shared understandings of forms of social life and social order' was *strengthened and made durable* once it became intertwined with localised contexts and, of course, individuals working within them (Jasanoff 2015a: 4). It is through this process of intertwining that we see how socially and politically acceptable knowledge of nuclear weapons was produced by civil defence leaders and volunteers at all levels of power, and then represented in British culture.[8]

This chapter examines how, at a regional level in 1950s Britain, a blend of nuclear sociotechnical imaginaries and localised context provided the rationale for civil defence exercises, parades and tournaments, and civil defence clubs: or, as Cronqvist (2015) has explored, these ritualised acts of civic duty. In the local and national press, the technology of nuclear weapons appeared benign or hidden through their imaginary alignment with familiar, safe and reassuring activities and shared values. The nuclear sociotechnical imaginary does not emerge as a 'static or tightly bounded belief system', but elements of this imaginary proved 'durable at the national level because [of] powerful instruments of meaning-making and goal-selecting [...] within the control of nation states' (Jasanoff and Kim 2009: 123). This speaks to 'embedding', which is concerned with the institutionalisation of an imaginary whereby 'the merely imagined is converted into the solidity of identities and the durability of routines and things' (Jasanoff 2015b: 323; see also Introduction, this volume). Anything discursive, hyperbolic or fictive created on nuclear policy by government that *did not* talk about the realities of nuclear weapons as unthinkably destructive weapons of war was part of a useful sociotechnical imaginary.

'Narrative Scenarios' of Nuclear Attack: Normalising Imaginary War

In the 1950s, the North Western Region Civil Defence Group, like all civil defence groups in the UK, performed various exercises to improve coordination and organisation, with the additional purpose of creating publicity for civil defence efforts. These exercises included elaborate introductory 'narratives' for the participants that imagined how the war had come about, described the resulting bomb damage, and outlined the likely consequences of this. These imaginary scenarios were commonplace and served to 'stabilise' the idea of survivable and manageable nuclear attack, thus performing a vision of acceptable nuclear use and promoting the viability of nuclear deterrence. They also represented a strand of denialism—a denial of reality in order to push away a psychologically uncomfortable truth—where undesirable future situations could be marshalled, controlled and made orderly. If denialism seems extreme and unlikely, then the idea of nuclear survivability can be made sense of by remembering the direct links that civil defence personnel habitually made to their experiences of civil defence work during the Second World War to the Cold War context.[9] In a short foreword to 'Exercise Miller', H. H. Schofield (Principle Officer, Home Office), said that 'to those who, like himself, served in Civil Defence during the last war he felt he need not here dilate on the importance of emergency feeding and improvisation'.[10] For many civil defence personnel who felt rightfully proud of their wartime service, Cold War civil defence was both a replication and extension of previous experience and knowledge, where survival and recovery were the achievable ultimate aims in a time before unpredictable radiation hazards. By the mid-1950s, official histories of civil defence were published and circulated to commemorate and celebrate pre-nuclear work.[11]

Many of the narrative scenarios deployed 'imagination' and 'reality' as active concepts to help frame or describe the exercises, and this language was echoed in local newspapers, demonstrating the construction and embedding of a stable sociotechnical imaginary that could be collectively understood. The scenarios tapped into the dramatic potential of the Cold War era, stressing the scale of nuclear devastation and the unthinkable catastrophe that could unfold. While these were, on the whole, sombre descriptions of the descent into World War Three, humour is occasionally detectable in the archival materials, suggesting some cynicism towards the Cold War situation more generally. In 1958, 'Exercise Joint Effort 2' was an exercise between eight different Joint, Civil and Military headquarters in the North-West region. The narrative begins in a familiar enough manner: 'The time is the winter of 196X [sic]', before turning to satire: 'The Eastern Bloc has achieved many further successes. For example two rhinoceros, bred in captivity, have just completed their millionth circuit of the earth in Sputnik 101'.[12]

More representative of the earnest approach to civil defence activities was the narrative scenario for 'Exercise Miller' in December 1952: '[I]t will be assumed that atomic bomb bursts have completely blocked both entrances of the Mersey Tunnel, severely damaged the underground railway station [...] many people are homeless'. A two-day conference was held in Birkenhead Town Hall on Merseyside, attended by around '80 local government, Home Office, Ministry of Food, Civil Defence and Women's Voluntary Service officials'.[13] The conference discussed how to feed citizens in the event of an atomic attack and 'Exercise Miller' was run on the Wirral, which led to specific sites being earmarked as emergency feeding centres. The conference required an exercise in imagination because the scenario involved the destruction of vital infrastructure. This nightmare scenario was represented in local newspapers, with headlines such as 'If an Atomic Bomb Fell on Birkenhead' or 'If A-Bombs Landed Here' or 'If Atom Bomb Hit Tunnel' accompanying reports on the conference.[14] This dramatic use of language sat alongside reassuring reminders that civil defence work was effective and increasingly well organised by devoted leaders and volunteers, while being crucial to ensuring the safety of British citizens in the nuclear age. Public discourse around civil defence generated the familiar representation of dutiful citizens supplementing the work of the state, offering a humane and altruistic face to nascent deterrence ideology, and a rejection of Cold War anxiety. These discursive and social constructs helped embed the nuclear sociotechnical imaginary at a local level and helped establish the 'real' and the 'imaginary' as familiar Cold War tropes.

The general awareness of the need to publicly imagine future nuclear catastrophe or come to terms with the uniquely powerful and dangerous properties of nuclear weapons was placed alongside the reassurance that 'real' civil defence work could allay fear.[15] Of course, this chapter argues that discursive construction of the 'real' was itself part of a social process that supported the sociotechnical imaginary of nuclear weapons. This preoccupation with both realism and imagination was explicit in civil defence exercises and their representation. For instance, a report on 'Exercise Wirral' in 1954 stated 'it was a difficult job to make an appreciation of what were to some extent imaginary incidents'.[16] On 15 March 1958, the *Birkenhead News* reported on 'Exercise Edinburgh' where 'many fires were raging in the area of operations and the cries of the injured and trapped victims could be heard above the roar of the flames [...] consisting of cases of fractured spines, severe head injuries, fractured legs, whilst in a smashed-in cellar they found and rescued victims with fractures of the arm or legs as well as those suffering from concussion and severe shock'.[17] This dramatic representation of the exercise was followed by reassurance to the reader about how well the highly trained civil defence personnel were able to deal with the situation. In a similar vein, but six years earlier, the *Birkenhead Advertiser* reported that as part of 'Exercise Morpeth' in 1952, 'for two hours nearly 500 volunteer firemen and Civil Defence workers, National Hospital Services Reserves and the W.V.S. dealt with imaginary incidents'.[18] Of course these incidents, and the scenarios, were imaginary, but the responses were real, and readers were invited to admire the personnel who could solve problems and survive. This discursive context downplayed the potential devastation of nuclear weapons, contained fear and promoted deterrence: this supported and embedded key aspects of the nuclear sociotechnical imagery and was a consistent discursive feature throughout the 1950s.

Also representative was the narrative used for 'Exercise Signal Fire', a fallout exercise in 1960. The exercise followed the apocalyptic sounding large-scale exercise 'Four Horsemen' in 1959, which involved 'Civil Defence Control, Signal, Scientific, Fire, Police, Hospital and Military Staff at Regional HQ and the Royal Observer Corps [...] grafted on a large-scale NATO air defence exercise'. The narrative for 'Signal Fire' was elaborate, detailing the context for worsening tensions between East and West, and then describing how the British public were being advised to 'take reasonable steps to preserve family life and to remain in their own homes'. The narrative ends on a dramatic note: 'Attack, if it comes, is expected to be from the east'. As the exercise unfolds, the report describes multiple '5-megaton' nuclear weapon air bursts and 'additional fallout hazard from bombs on the Continent and on the south of the United Kingdom'.[19] It was assumed that people would evacuate to Cheshire from Liverpool through the Mersey tunnel. 'Exercise Mermaid II', held on 27 May 1962, offered a blander description: '[I]n the early hours of this morning widespread nuclear attacks have taken place on this country'.[20]

The juxtaposition was constant throughout the early Cold War era—reassuring rhetoric and activities sat alongside actively imagining nuclear attack. This also characterised *representations* of civil defence, and this discursive context should be seen as an important aspect of the sociotechnical imaginary that was generated to support deterrence. For instance, the recruitment film *The Waking Point* (1951), screened in local cinemas throughout the 1950s, encapsulated this perfectly: nuclear fear was represented as an inevitable part of life, but joining civil defence—ideally as a citizen who conformed to the gendered behaviour and thinking appropriate to the Cold War era—was presented as an active way to counteract it (Hogg 2016: 81–82; Grant 2013: 11–12). This brief example shows how ideas and particular framings of real and imagined nuclear danger were appropriated for propaganda purposes in the name of nuclear deterrence.[21] In the 1950s, fictional nuclear attack scenarios were a common feature in popular culture and the national and local press, which created an influential discursive context that informed the articulation of civil defence narrative scenarios.[22]

Once the scene was set and exercises were completed, they were evaluated. 'Exercise Miller' was deemed a huge success because disparate groups within government had worked together effectively to plan how to provide food for a dispersed population. At its simplest, improvised ovens and campfires would be provided at large outdoor sites, a clear throwback to Second World War-style organisation. The huge challenges involved, including predicting where (or advising how) people would congregate, ensuring secure and plentiful food supplies for a fixed period of time, loomed larger than considerations of radiation danger or acknowledging the possible chaos of a post-nuclear scenario. Press reports focused on the excellent levels of organisation and competence displayed by (predominantly male) government officials taking part in these imaginary scenarios. As with many civil defence exercises in the 1950s, the public representation of nuclear war was both dramatic and controllable.

The narrative scenarios used to frame civil defence and then the localised responses to civil defence exercises produced socially and politically acceptable knowledge of nuclear weapons, which both normalised a postwar nuclear 'peace' and reinforced aspects of the dominant sociotechnical imaginary. The promotion of imagined nuclear survivability as a possible future gave the impression that there were no insurmountable problems for civil defence volunteers and their leaders. Exercise narratives and the way the exercises were reported after the event assumed an orderliness and predictability, and the archival record holds few echoes of emotional responses, or the likelihood of human unpredictability.

Civic Engagement and Sites of Leisure

As part of the process of implementing nuclear civil defence strategy, various sites of leisure (such as social clubs) were created and activities (such as quizzes, motor rallies or regional tournaments) were introduced to promote participation, entrench membership and community belonging, and to generate positive publicity. Although small in scale, these practices fed into similar efforts on a national scale and promoted values that helped embed the nuclear sociotechnical imaginary (such as militarism and duty) through 'ritual' and spectacle (Kong and Yeoh 1997; MacDonald 2006). The materiality, social organisation and physical institutionalisation of civil defence on a local level promoted deterrence ideology by representing the nation's military nuclear technology project as peaceful and normal.

Matthew Grant has argued that civil defence 'from 1951 was represented as inclusive, providing associational opportunities and leisure satisfaction', but as the Cold War progressed 'the culture of civil defence, and the assumptions of political culture concerning its popular appeal, were increasingly divorced from the realities of British culture' (Grant

2011: 54, 60). In the 1950s however, 'the core message of participation in national defence was adapted, stressing "patriotism" less and emphasizing that to serve in civil defence was to fight for peace and to protect the local community' (Grant 2011: 54). Notions of participation and duty were central ideas to encourage and rework in the nuclear age. As Miss Hornsby-Smith said in her speech that concluded the North-West regional 'Tourney' in 1958: 'Today you have given us a display of skill and techniques which has earned well-merited applause; but you have also the deeper satisfaction of performing a public service of the highest order.'[23] The results of the Tourney were sent to 42 local newspapers, demonstrating the public relations objectives of these tournaments. Serious effort went into organising regional civil defence tournaments, mainly because they were viewed as good publicity for civil defence initiatives: Hornsby-Smith argued that 'we have got to "sell" civil defence'.[24] The tournaments comprised of exhibitions and demonstrations, as well as competitions between civil defence groups in the region.[25] On successful completion of the tournaments, congratulations from national leaders and dignitaries (such as the Duke of Gloucester in 1959) always followed, which suggests the level of prestige and importance attached to the success of these meetings, for both morale purposes and publicity.[26]

Civil defence groups were an active part of communities, would attend commemorative events such as Remembrance Sunday, and take part in civic parades.[27] In a more durable social sense, a physical 'Civil Defence Club' was created in Birkenhead in 1956, offering an indication of the way in which civil defence practices and personnel became embedded into the fabric of everyday life in the North-West region. The archives hold plenty of detail on the plans to create a club that would allow volunteers and visiting civil defence personnel to enjoy the (then) overwhelmingly masculine pursuit of playing billiards, drinking at the bar, and trying out Home Office financed luxury wicker seats and tables.[28] These mundane developments gently interwove normative imaginaries around leisure pursuits with nuclear civil defence aims. As with the discursive context we saw in the previous section, these efforts socialised and normalised soft militarisation in the postwar era as part of the modernising project of nuclear deterrence. One example of how this militarisation became more visible was the use and representation of Cold War machinery. In 1955, a civil defence helicopter toured UK civil defence groups to demonstrate how radiation might be measured.[29] By exhibiting reassuring, new and virtuous technologies capable of controlling risk, and therefore containing fear, the sight and sound of a helicopter overhead aided the normalisation of the peaceful nuclear weapons imaginary.

As part of the discursivity of 'embedding' ideas that supported civil defence socialisation, regional groups regularly received civil defence information and propaganda films from central government. The American documentary film, *Survival Town* (1955) was shown in cinemas, and was accompanied with civil defence promotional material.[30] An official civil defence circular received by the Birkenhead town clerk in 1956 described the effects of 'new weapons' and the importance of a new imaginary: 'an attack with new weapons [...] would be a struggle for survival of the grimmest kind'.[31] In the same year the Mayor of Birkenhead and other local dignitaries received an invitation from Whitehall to view the premiere of the government produced film *The Hydrogen Bomb* (1956) at the Walker Art Gallery, Liverpool.[32] The film begins with the slow, solemn beat of a drum, matched by the funereal and foreboding tone of the narrator. Warning the audience of 'this new horror' he reminds everyone that 'we must prepare ourselves, every one of us' and that if nuclear war happens, 'we must accept the fact that we shall all be in it'. Touching upon the shared cultural memory of the Second World War, the narrator states that, 'as was experienced in the last war', calls for help were never left unanswered. Presumably this national character would live on and defines the civil defence volunteers who are ready to help in the event of nuclear war. The point is reiterated: 'when peace returned to our islands, we had hardly taken stock of our resources when we were confronted with a new danger more frightful than ever before [...] gradually our minds began to accept this horror [...] are we to sit back complacently and say "bad luck Jack, I'm alright"'.[33] The weapons were presented as inevitable, but recast as manageable and passively defensive. Yet, the response of British citizens needed to be active: good character, duty and responsibility need to be demonstrated in the nuclear age. The initial introduction of fear as an understandable reaction to the thermonuclear era is systematically rejected. The familiar visual vocabulary of concentric circles, blast effects and ineffective citizens was used to promote civil defence aims.[34] These strategies and imaginaries were consistent: one film from the year before, *Civil Defence Shows Its Paces* (1955), claimed that 'enough of us can remember the real thing to hope that the far greater horror of nuclear war will never become a reality'.[35]

With the advent of the H bomb, central government advised that 'officers must think about vast areas of damage, and in terms of homeless amounting to hundreds of thousands [...] the accent must be on planning in the future and not upon training as in the past [...] all future planning must be based upon dispersal, mobility and maintenance of communications'.[36] Additionally, *The Hydrogen Bomb* (1957), a government produced civil defence pamphlet, reinforced the importance of public acceptance of deterrence and argued that dissent 'might have serious, or even fatal, consequences'.[37]

Leaders in the region were being encouraged by national government to coordinate new types of civic engagement and information dissemination in the nuclear age. The communal viewing of *The Hydrogen Bomb* in the grand setting of the Walker Art Gallery is a good example of how the sociotechnical imaginary of nuclear weapons became embedded in this localised social context. The use of existing sites of leisure and culture was subtly transformed for Cold War purposes, and new permanent sites were created. The intertwining social context and physical enactment on the local level were enabled and made durable through assumptions that were reiterated through communication with national leadership, and also echoed in the national press. After exploring sites of leisure, civic engagement and associated propaganda, we can conclude that civil defence personnel were simultaneously producing their own collective civic identity and embedding the nuclear sociotechnical imaginary in society.

Imaginative Geographies

The final way in which we might unpack the nuclear sociotechnical imaginary as 'imaginative infrastructure' is to examine the localised geographical context of civil defence work on the Wirral peninsula. I argue that imaginative geographies helped create the illusion of safety, control and predictability in local civil defence planning, and also particular representations of airspace promoted deterrence ideology and the development of military nuclear technology as national virtues.

The exercise scenarios analysed earlier often deployed, and depended upon, an imagined nuclear geography. In 'Exercise Miller', references to geographical context shaped the imagined scenario and the exercise that followed in important ways. Mr. Pugsley, the Regional Food Officer stated that 'because of its geography, Wirral had its own particular problems that had to be faced', with particular reference to the likely destruction or blockage of the Mersey tunnel, and the fact that the population would be surrounded by water on three sides.[38] Yet, the materiality of the Wirral peninsula apparently made the movement of people more predictable in the event of nuclear war. It was assumed that tens of thousands of people from Birkenhead and Wallasey on the banks of the Mersey would move west into 'open country' or parkland. People would most likely eventually travel south-west towards the more 'remote' region of North Wales, deemed safer due to the lack of industrial targets. The focus of civil defence planners was on containing and controlling large groups of people in open spaces where possible 'who might otherwise restrict movement of incoming forces who are known to be on their way'.[39] Interestingly, there is no discussion of northward evacuation by boat or ship, which would doubtless be an instinctive move by a proportion of the population of Merseyside with seafaring ability and knowledge, but an obvious potential flashpoint of disorder if masses of people attempted to board boats destined for the open sea, The Isle of Man or Ireland. Nowhere is the likely scenario discussed that burn victims might look to the treacherous tidal waters of River Mersey for relief, or that, with good knowledge of the tide, it is possible to walk and swim to North Wales via the River Dee on the west side of the Wirral peninsula. Knowledge of the sea could prove an invaluable skill in a post-apocalyptic world. Of course, these are my own speculative attempts to create imaginary geographies of unpredictable post-nuclear reality. The archival record reflects a more rigid control of the parameters of nuclear emergency.

Perhaps the geography of the peninsula offered planners a degree of overconfidence in predicting how people would react in the event of atomic war, when in actual fact peninsula geography could lead to especially unpredictable and perilous patterns of behaviour in the days and weeks after a nuclear strike. The imaginative geographies in the archive are limited in scope and of course reflect the desire to achieve the practical aims of the exercises. Yet, these assumptions about nuclear emergency reflect confidence in survivability and necessarily downplay nuclear harm. In this sense, it was clearly important to control the fictive elements of civil defence work and thus define the imaginative boundaries for civil defence personnel.

Following Edward Said's formulation of the concept (Said 1978), Matthew Farish describes imaginative geographies as 'influential presentations of the world and its contours that are made possible by particular forms of knowledge about that world. However propagandistic or fantastic they might seem [they] are only fictional in the sense that they have been fabricated. With sufficient circulation, or attendance, they could alter attitudes or policies' (Farish 2010: xii). In a similar style to the concept of sociotechnical imaginaries, these constructs have the potential to be powerful social agents. The creation of imaginative geographies in civil defence work activated reassuring and non-problematic spaces, therefore playing a role in embedding the three elements of the nuclear sociotechnical imaginary introduced in this chapter.

As we saw in 'Exercise Morpeth', jet planes were a new feature in British skies in the postwar era. As well as a potentially dangerous physical space containing jets, missiles, nuclear weapons and military personnel, Cold War airspace was represented as a space of vulnerability (Hogg 2016). Yet, particular representations of airspace were crucial to the promotion of deterrence ideology and the development of military nuclear technology as a 'safe' national virtue. Representations of new nuclear 'V-bombers' in the national press, Pathé newsreels and popular science magazines contributed to the soft propaganda surrounding Cold War militarisation.[40] The increasing popularity of airshows, those jingoistic jamborees of the nuclear age, reflected people's fascination with ultra-modern aircraft like the Avro Vulcan. These futuristic machines effortlessly reflected British engineering ingenuity, and their public appearances served to promote and normalise the militarised deterrence posture that was increasingly dependent on the complete control of airspace.[41] This element of the sociotechnical imaginary was useful in counteracting discourses of nuclear threat and fear, influenced perceptions of nuclear danger, and was capable of contributing to a positive Cold War spectacle.

Conceptions of airspace proved to be a powerful part of the reorganisation of the relationship between citizens and state that the Cold War entailed. In the Cold War era, an unusually high percentage of land in the UK (1.5%) was eventually assigned to all military practices in the Cold War (Williams 2011; Flintham 2010). As jet and missile technologies developed, the British nuclear state needed to inhabit, police and regulate airspace. Thus huge defensive installations (such as Fylingdales) were developed to detect incoming airborne threats alongside the aggressive machinery created to 'preserve peace' (Edgerton 2013). David Edgerton argues that a myth of 'niceness and decency', with its roots in the nineteenth century, meant that the UK did not foster a tradition of 'serious analysis or critique of technology', so the scale and persistence of the 'warfare state' became more-or-less invisible in the postwar era. Defined and policed airspace had become a normal method of territorial control by the end of the Second World War. Concentrating on military practices, Flintham views airspace as 'partly organized around the vulnerabilities of the human civilian body' (Flintham 2010: 14). Airspace is defined by varieties of control, where the space 'project[s] its many lethal forms across the landscape and into the skies and fashion[s] the most convoluted structures to protect and deliberately restrict the movements of a civilian community' (Flintham 2010: 14). So, airspace can be a vital controlling mechanism as well as a vulnerable space.

Cold War airspace was the space through which missiles or jet planes would travel and find their targets and where radioactive fallout could travel unpredictably on the wind. In March 1958 Exercise Edinburgh was carried out in Merseyside, and its success was proclaimed in *Birkenhead News* with the headline "A" Bomb Exercise was Staged in Birkenhead. The main report said, 'an imaginary "Inter Continental Ballistic Missile" landed on Merseyside at 8.30 a.m. last Sunday, bring[ing] the elements of nuclear warfare to the Birkenhead area where "damage" ranged from total destruction in some quarters to only slight damage in others'. The newspaper extended the illusion of attack and included gruesome details of imagined injuries. Yet, the official exercise narrative included the sentence 'no problem of fallout is set in this exercise', and there was no mention of radiation to the readership of the local press beyond the mention of 'A-bomb' in the headline. The public were reminded once again of the imaginary potential of nuclear attack from the sky, but not the radiological after-effects of nuclear weapons.[42] This again demonstrates that when exercises were made manageable in practical terms, the dangers of nuclear war were routinely compartmentalised and sanitised to accommodate the fictive sociotechnical imaginary. These spatialised imaginaries once again reveal how the limits placed on the Cold War imagination aided the promotion of particular ideas and attitudes.[43] As Matthew Farish suggests of the Cold War:

> [A]t no other juncture in history had global, regional, continental, and urban spaces been wrapped into a single "regime of truth" that delineated the parameters of reliable knowledge in the antagonistic and divisive terms of national strategy and in turn defined strategic knowledge literally along specific geographic lines. (Farish 2010: xvii, xiii)

Nuclear sociotechnical imaginaries not only transformed how British citizens understood urban and regional space, but also functioned to control those spaces.

Conclusion

This chapter has explored civil defence culture from the perspective of the regional leadership tasked with implementing national policy. Faced with the unthinkable consequences of nuclear attack, regional civil defence leaders interpreted the exercises they were tasked with through the lens of their social and geographical context and the limited horizons of the official nuclear imagination.

Seen in the light of normative assumptions about the danger of nuclear war, the narrative scenarios that accompanied the civil defence exercises can appear uncreative and repetitive and can conform to the limitations imposed by the processes of the nuclear sociotechnical imaginary. The scenarios reinforced existing Cold War imaginaries (control, predictability, permanence of the bomb) through controlling the fictive elements of the exercises, meaning that imagined disaster conformed to a normalised and sanitised vision of nuclear war. Thus, the sociotechnical imaginary, at least through the way civil defence exercises were conceived, remained discursively familiar and predictable.

Although beyond the scope of this chapter, international Cold War research suggests that we can detect similarities in how imaginaries translated into civil defence practice in different national contexts. Observations in recent research on 1950s Denmark (Farbøl 2020) suggest that parallels might be drawn when looking at how national policy was translated to localised contexts, particularly when looking at how civil defence leadership attempted to achieve the impression of realism during exercises. It is hoped that this chapter has offered one way to conceptualise the local–national–international analytical frame in relation to sociotechnical imaginaries.

The chapter also suggested that powerful discourse reinforced the sociotechnical imaginary, implying that individuals would gravitate towards supporting these imaginaries. It is necessary, though, to remind ourselves that civil defence personnel would have had a variety of feelings and attitudes towards their Cold War roles. As Douthwaite states:

> [T]he environments in which civil defence recruits embarked on training were mixed and contested in meaning and purpose. Recruits were obliged to understand and engage with several different nuclear experiences at once by training to protect the public in a real nuclear attack, enacting their symbolic roles in deterrence strategy, and navigating multiplying and contradictory narratives of deterrence policy and nuclear vulnerability in the public sphere. (Douthwaite 2018: 180)

When discussing discourse or sociotechnical imaginaries, ignoring individual agency is a danger. Although this chapter has not focused much on the agency of civil defence personnel, it is hoped that it has offered some ideas on how we might historicise the structural role and function of imaginaries.

We know that the key aspects of the sociotechnical imaginaries described here came under increased scrutiny as the 1950s progressed, not only by anti-nuclear activists but also by regional leaders tasked with implementing civil defence policy as well (Barnett 2015). While this chapter has historicised the influence that imaginaries had, it has also highlighted some of the fragile and contradictory social elements at play at the local level which would be lost with a sole focus on the national picture.[44]

Reflecting one last time on the work of Sheila Jasanoff and Sang-Hyun Kim, in this chapter I have attempted to show that the nuclear sociotechnical imaginary that emerged around civil defence in the UK in the 1950s did 'describe attainable futures and prescribe[d] futures that states believe ought to be attained' (Jasanoff and Kim 2009: 120). The case studies at the heart of my analysis suggest that imaginaries 'and the policies built upon them, ha[d] the power to influence technological design, channel public expenditures, and justify the inclusion or exclusion of citizens with respect to the benefits of technological progress' (Jasanoff and Kim 2009: 120). Ideas, imaginaries and illusions perpetuated the nuclear state and allowed deterrence ideology to remain legitimate, because 'in activating collective consciousness, imaginaries help[ed] create the political will or public resolve to attain them' (Jasanoff and Kim 2009: 123). What was attained in real terms (the desirable utopian future) was the continued existence of nuclear weapons and all the institutional, scientific and technological apparatus that is required to enable and contain this. A detailed examination of the localised social, discursive and geographical contexts within which civil defence activism occurred has allowed this chapter to identify and understand the ways in which these nuclear sociotechnical imaginaries became embedded in British life. Perhaps it is the improvised oven used in the atomic age as part of 'Exercise Miller' where we see 'imagination, objects and social norms – including accepted modes of public reasoning and new technological regimes – becom[ing] fused in practice' (Jasanoff 2015b: 322).

Notes

1. 'Jets Made Biggest Dock Raid Since the War', *Birkenhead News* 9 April 1952.
2. Wirral Archives (WA) B/165/13 Birkenhead Fire Brigade: Exercise Morpeth, March/April 1952.
3. A historiographical discussion follows, but the layered definition of 'sociotechnical imaginary' outlined in Jasanoff (2015b) has inspired the approach taken in this chapter.
4. The discursive context is particularly important in helping us to understand how and why ideas about nuclear weapons were generated, discussed and reinforced on a daily basis. 'Discursive', in the context of this chapter, is understood as the set of practices that enable public dissemination of discourse surrounding nuclear weapons. This suggests a relationship between the public expression of knowledge (through discursive constructions) and power (the ability to sustain nuclear weapons deployment). Clearly, this formulation follows Michel Foucault's concept of discourse and the 'discursive field' first described in his 1970 lecture 'The Order of Discourse' (Young 1981) and widely adopted by scholars in the last few decades. I hope to show how discursive practices were central to the creation of a powerful sociotechnical imaginary of nuclear weapons.
5. This blurring of the lines between nuclear and conventional civil defence was clearly a transnational phenomenon: see Farbøl (Chapter 7), Molitor (Chapter 3), and Bennesved and Sylvest (Chapter 5) in this volume.
6. See Chapter 4 in Campbell (1983) for Civil Defence Corps recruitment numbers. Estimates replicated elsewhere are 300,000 + by late 1953, rising to 375,000 + by 1963, not including at least another 200,000 members in other Civil Defence organisations.
7. See the chapter by Bennesved and Sylvest (Chapter 5) in this volume.

8. See also the chapter by Cronqvist and Grant (Chapter 9) in this volume.
9. See also the chapter by Cronqvist and Grant (Chapter 9) in this volume.
10. WA B/165/6/3 Home Office—Ministry of Food: North Western Region: Exercise Miller: Civil Defence—Emergency Feeding, December 1952, 1.
11. Institutional memory of wartime service was generated and then linked to civil defence efforts in the 1950s. An official history, *Civil Defence* by T. H. O'Brien, published by HMSO as part of their series of 'Civil Histories of the Second World War' was circulated to regional civil defence leadership in 1955. Regional leaders were encouraged to bring the publication to the attention of members and officials in the local authority. WA B/165/9/1 Letter from Director of Publications, HMSO to Birkenhead Town Clerk, November 1955.
12. WA B/165/13 Exercise Joint Effort II: Narrative, 1958.
13. WA B/165/6/3 *Birkenhead Advertiser*, 17 December 1952.
14. WA B/165/6/3 Civil Defence: Food Ministry Emergency Feeding Exercises, 1952.
15. See also Bjørnsson (Chapter 2) in this volume on the reassuring nature of action.
16. WA B/165/13 Experimental Mobile Column: Exercise Wirral: Report of Post-exercise conference, 1954.
17. *Birkenhead News*, 15 March 1958.
18. *Birkenhead Advertiser*, 9 April 1952.
19. WA B/165/14/1 North Western Region report on Exercise 'Signal Fire', 9 and 10 April 1960.
20. WA B/165/17/1 Exercise 'Mermaid II', 27 May 1962.
21. See Grant (2010) and Hogg (2016) for discussion of such material.
22. The genealogy of journalistic use of fictional scenarios would benefit from sustained collation and analysis. There is certainly a case to be made for a consistent set of assumptions across time about the knowledge base of the 'ideal reader', which led to certain aspects of an attack scenario and its consequences not being explained in detail.
23. WA B/165/13/3 Miss Hornsby-Smith's Speech at the North-Western Regional Tourney, 7 June 1958.

24. WA B/165/13/3 Miss Hornsby-Smith's Speech at the North-Western Regional Tourney, 7 June 1958.
25. WA B/165/13/1 shows the significant organisational effort that went into these events.
26. WA B/165/13/3 Letter from Lieutenant General E. N. Goddard to P. Heath, Town Clerk, 15 July 1959.
27. WA B/165/9 Remembrance Sunday Ceremony in London, 10 November 1957.
28. WA B/165/9 Civil Defence and Civic Club Inventory, June 1957.
29. *The Northern Whig and Belfast Post*, 15 September 1955: 5.
30. WA B165/9/1 Letter from the Regional Director of Civil Defence to the Town Clerk of Birkenhead, 27 October 1955.
31. WA B165/9/1 Home Office, Civil Defence Circular No. 28/1956, 27 September 1956.
32. WA B165/9/1 Letter from the Mayor of Birkenhead to Birkenhead Town Clerk, 15 June 1956.
33. WA B165/9/1 'The H-Bomb: Film Commentary As Recorded' contained with correspondence dated 5 June 1956.
34. See Bennesved and Sylvest (Chapter 5) in this volume for more on the obfuscation of the political nature of nuclear war or how the nature of nuclear war was underspecified more generally.
35. Her Majesty's Stationery Office (HMSO), *Civil Defence Shows Its Paces* (1955).
36. WA B165/9/1 Notes taken at a meeting of the heads of civil defence services, 21 February 1957.
37. HMSO, *The Hydrogen Bomb* (UK: HMSO, 1957), 22.
38. WA B/165/6/3 'They Plan Feeding For 400,000 If A-Bomb Drops Here', *Birkenhead Advertiser*, 17 December 1952.
39. WA B/165/6/3 Home Office – Ministry of Food: North Western Region: Exercise Miller: Civil Defence—Emergency Feeding, December 1952: 46.
40. *Air News: The Vulcan* (1953). Available at https://www.britishpathe.com/video/air-news-the-vulcan/query/VULCANS [accessed 15 March 2021].
41. On 'machine dreams' see Fritzsche (1992), and for the broader political cultures of the air, see Adey (2010).
42. *Birkenhead News* 15 March 1958; WA 165/17/7 Exercise Edinburgh 'Narrative', 1958.

43. This was also echoed in national discourse, strengthening links between local and national assumptions over nuclear danger (Hogg 2016).
44. See also Cronqvist and Grant (Chapter 9) in this volume on the notion of 'fuzziness' in relation to sociotechnical imaginaries.

References

Unpublished Sources

Wirral Archives, UK.

Literature

Adey, Peter. 2010. *Aerial life: Spaces, mobilities, affects*. London: Wiley-Blackwell.
Alexis-Martin, Becky, and Thom Davies. 2017. Towards nuclear geography: Zones, bodies, and communities. *Geography Compass* 11: 1–13.
Anderson, Benedict. 1983. *Imagined communities: Reflections on the origin and spread of nationalism*. London: Verso.
Barnett, Nicholas. 2015. 'No protection against the H-Bomb': Press and popular reactions to the coventry civil defence controversy, 1954. *Cold War History* 15: 277–300.
Campbell, Duncan. 1983. *War Plan UK*. London: Burnett.
Cronqvist, Marie. 2015. Evacuation as welfare ritual: Cold War media and the Swedish culture of civil defence. In *Nordic Cold War cultures: Ideological promotion, public reception, and East-West interactions*, eds Valur Ingimundarson and Rosa Magnusdottir, 75–95. Helsinki: Aleksanteri Cold War Series.
Douthwaite, Jessica. 2018. *Voices of the Cold War in Britain, 1945–1962*. PhD Thesis. University of Strathclyde.
Edgerton, David. 2013. *England and the aeroplane: Militarism, modernity and machines*. London: Penguin.
Farbøl, Rosanna. 2020. Urban civil defence: Imagining, constructing and performing nuclear war in Aarhus. *Urban History* 1–23.
Farish, Matthew. 2010. *The contours of America's Cold War*. Minneapolis, MN: University of Minnesota Press.
Flintham, Matthew. 2010. The Shoeburyness Complex: Military space and the problem of the civilian body. In *Militarized landscapes: From Gettysburg to Salisbury Plain*, eds Chris Pearson, Peter Coates and Time Cole, 81–94. London: Continuum.

Fritzsche, Peter. 1992. *A nation of flyers: German aviation and the popular imagination*. London: Harvard University Press.
Grant, Matthew. 2010. *After the bomb: Civil defence and nuclear war in Britain, 1945–68*. London: Palgrave Macmillan.
Grant, Matthew. 2011. 'Civil defence gives meaning to your leisure': Citizenship, participation, and cultural change in Cold War recruitment propaganda, 1949–1954. *Twentieth Century British History* 22: 52–78.
Grant, Matthew. 2013. Images of survival, stories of destruction: Nuclear war on British screens from 1945 to the early 1960s. *Journal of British Cinema and Television* 10(1): 7–26.
Gregory, Derek. 1995. Imaginative geographies. *Progress in Human Geography* 19: 447–485.
Hill, Christopher R. 2018. *Peace and power in cold war Britain: Media, movements and democracy, c.1945-68*. London: Bloomsbury Academic.
Hogg, Jonathan. 2012. 'The family that feared tomorrow': British nuclear culture and individual experience in the late 1950s. *The British Journal for the History of Science* 45: 535–549.
Hogg, Jonathan. 2016. *British nuclear culture: Official and unofficial narratives in the long twentieth century*. London: Bloomsbury Academic.
Hogg, Jonathan, and Kate Brown. 2019. Introduction: Social and cultural histories of British nuclear mobilization since 1945. *Contemporary British History* 33: 161–169.
Hughes, Jeff. 2003. The Strath Report: Britain confronts the H-Bomb, 1954–1955. *History and Technology* 19: 257–275.
Jasanoff, Sheila. 2015a. Future imperfect: Science, technology and the imaginations of modernity. In *Dreamscapes of modernity: Sociotechnical imaginaries and the fabrication of power*, eds. Sheila Jasanoff and Kim, Sang-Huyn, 1–33. Chicago: Chicago University Press.
Jasanoff, Sheila. 2015b. Imagined and invented worlds. In *Dreamscapes of modernity: Sociotechnical imaginaries and the fabrication of power*, eds Sheila Jasanoff and Sang-Huyn Kim, 321–41. Chicago: Chicago University Press.
Jasanoff, Sheila, and Kim, Sang-Huyn. 2009. Containing the atom: Sociotechnical imaginaries and nuclear power in the United States and South Korea. *Minerva* 47: 119–146.
Jasanoff, Sheila, and Kim, Sang-Huyn, eds. 2015. *Dreamscapes of modernity: Sociotechnical imaginaries and the fabrication of power*. Chicago: Chicago University Press.
Kalmbach, Karena, Andreas Marklund, and Anna Åberg. 2020. Crises and technological futures: Experiences, emotion, and action. *Technology and Culture* 61: 272–281.

Kong, Lily, and Brenda S. A. Yeoh. 1997. The construction of national identity through the production of ritual and spectacle. *Political Geography* 16: 213–239.

MacDonald, Fraser. 2006. Geopolitics and 'the vision thing': Regarding Britain and America's first nuclear missile. *Transactions of the Institute of British Geographers* 31: 53–71.

Masco, Joseph. 2008. 'Survival is your business': Engineering ruins and affect in nuclear America. *Cultural Anthropology* 23: 361–398.

Monteyne, David. 2011. *Fallout shelter: Designing for civil defense in the Cold War*. Minnesota: University of Minnesota Press.

Noakes, Lucy, and Juliette Pattinson, eds. 2013. *British cultural memory and the Second World War*. London: Bloomsbury Academic.

Polleri, Maxime. 2020. Post-political uncertainties: Governing nuclear controversies in post-Fukushima Japan. *Social Studies of Science* 50: 567–588.

Rech, Matthew F. 2015. A critical geography of observant practice at British military airshows. *Transactions of the Institute of British Geographers* 40: 536–548.

Robinson, Sam. 2020. Scientific imaginaries and science diplomacy: The case of ocean exploitation. *Centaurus*, 1–21.

Shaw, Tony. 2005. Introduction: Britain and the cultural cold war. *Contemporary British History* 19(2): 109–115.

Said, Edward. 1978. *Orientalism*. New York: Pantheon Books.

Sismondo, Sergio. 2020. Sociotechnical imaginaries: An accidental themed issue. *Social Studies of Science* 50: 505–507.

Williams, Alison J. 2011. Reconceptualising the spaces of the air: Performing the multiple spatialities of UK military airspaces. *Transactions of the Institute of British Geographers* 36: 253–267.

Young, Robert J. C., ed. 1981. *Untying the text: A post-structuralist reader*. Boston: Routledge.

Zweiniger-Bargielowska, Ina. 2000. *Austerity in Britain: Rationing, controls and consumption, 1939–1955*. Oxford: Oxford University Press.

CHAPTER 5

Embedding Preparedness, Assigning Responsibility: The Role of Film in Sociotechnical Imaginaries of Civil Defence

Peter Bennesved and Casper Sylvest

By the late 1950s and early 1960s, it was common for anti-nuclear campaigners and government officials alike to complain about public apathy towards the nuclear threat. Explanations offered for this nuclear withdrawal displayed some similarities, although their implications were poles apart. Public intellectuals like Günther Anders and Lewis Mumford essentially faulted the alienating dynamics of modernity and capitalism (van Munster and Sylvest 2016), whereas a 1962 US National Research

P. Bennesved (✉)
Department of Historical, Philosophical and Religious Studies, University of Umeå, Umeå, Sweden
e-mail: peter.bennesved@foi.se

C. Sylvest
Department of History, University of Southern Denmark, Odense, Denmark
e-mail: csy@sdu.dk

M. Cronqvist et al. (eds.), *Cold War Civil Defence in Western Europe*,
https://doi.org/10.1007/978-3-030-84281-9_5

Council report on *Behavioral science and civil defense* suggested that widespread anxiety led to an unwillingness to confront difficult questions (Baker and Cottrell 1962: 75–76). Commenting on the latter study, Spencer Weart concluded in the mid-1980s that 'up to a point, most people could understand what war might really mean', but soon after 'the imagination rebelled' (Weart 1987: 20). In sociotechnical imaginaries of civil defence during the Cold War, the role of film was to assist the imagination in order to recruit and educate and to do so in a manner that provided hope and reassurance and served the political objectives with which civil defence organisations were invested. Visual media were central to public information campaigns at a time when cinema audiences were substantial and television began to make inroads in Western societies. Expanding state apparatuses had already exploited the potential of film during the Second World War. In the atomic age, inaugurated in popular consciousness by the bombings of Hiroshima and Nagasaki in August 1945, state-sponsored film found a new territory.

The study of civil defence films has had an American tilt, which reflects the central standing of the USA at the time and the postwar dynamics of Americanisation. Their differences notwithstanding, American films rehearsed a familiar theme: something could be done, and preparation was of the essence (e.g. Masco 2008; Jacobs 2010; Matthews 2012; Nowak 2016). Historians have initiated studies of civil defence films in other parts of the Cold War West (e.g. Cronqvist 2008; Norén 2012; Grant 2013; Bennesved and Norén 2020), and scholarship in this area carries significant potential for deepening our understanding of the similarities and differences of national civil defence efforts. Neutral Sweden and NATO-allied Denmark differed significantly in terms of civil defence in the postwar decades.[1] Still, their close relationship in the immediate postwar years and their increasingly divergent positions in the Cold War present an opportunity to break out of national narratives. Like the technology that came to define it in the nuclear age, civil defence and its associated cultures were transnational phenomena (also Berger et al. 2018). Against this background, it is the ambition of this chapter to place the study of civil defence films in a transnational framework and to initiate efforts of comparison through a theoretically informed study of films produced in Sweden and Denmark during the early nuclear age (c. 1949–1965). We conceptualise civil defence as a sociotechnical imaginary and focus on the role of film in processes of embedding during the early nuclear age. We ask: What was the function of film in these two countries in embedding an imaginary of civil defence before and during the thermonuclear revolution? And how was the film medium exploited to structure the imagination of nuclear war and ways of surviving such a catastrophe?

Underlying our analytical strategy and the decision to analyse early (c. 1949–1954) and later films (c. 1954–1965) in this period is a historically informed assumption revolving around the development of thermonuclear weapons technology: though irregular and occasionally unsystematic, the general drift of civil defence thinking during the late 1950s was towards increasing individualisation of responsibility.[2] Whereas early postwar civil defence witnessed much continuity with the concept of aerial protection—signalling an uneven adjustment to the risk of atomic warfare—the mid-1950s was marked by new, fraught questions about the nature of H bomb warfare. What protective measures were available against the immediate and longer-term effects of these unfathomable bombs at a time when new means of delivery were clearly on the horizon—and were available measures credible? Such questions became central to civil defence. Over time, they gave rise to new policies. In this process, collectively organised civil defence efforts and notions of civilian preparedness and responsibility were transformed. This transformation appears to have been central to the subsequent reshaping of civil defence in the 1970s and 1980s, yet our main focus is on determining whether, how and with what consequences this trend is reflected in filmmaking of the early nuclear age.

Sociotechnical Imaginaries and Visual History: Civil Defence Films as Practices of Embedding

The concept of sociotechnical imaginaries is promising for advancing civil defence history (see Introduction, this volume; Jasanoff and Kim 2015; Jasanoff 2015a). While civil defence in Sweden and Denmark can be analytically approached as imaginaries in their own right, it is also apparent that these organisations played a part in upholding a wider vision of stability and progress, in part by containing and curbing the considerable risks of Cold War security politics. In this section, we attune the concept of sociotechnical imaginaries to the study of films and television. Though visual material took many forms and was used for a host of purposes, we focus on moving imagery about civil defence produced by state authorities and their associated partners intended for education, instruction, recruitment and propaganda. We zoom in on one particular dimension of such imaginaries: the study of intended use of visual media for embedding purposes, which we conceive as a combination of material structures and artefacts, a thoroughly technologised set of activities and a state of mind. We argue that the study of films harbours a potential to harness and refine the approach set out by Jasanoff and Kim (2015). We make three theoretical contributions.

First, a significant part of the conscious effort to embed postwar imaginaries of civil defence in Sweden and Denmark was centred on film, a potent medium for recruitment and information aimed at large audiences. Jasanoff develops the concept of embedding primarily with reference to the 'production of things' or collective forms of 'remembering' that transform 'the merely imagined into the solidity of identities and the durability of routines and things' (Jasanoff 2015b: 329, 323). While she places considerable theoretical emphasis on ways of seeing and the role of perception (Jasanoff 2015a: 13–14), she does not explicitly mention film among psychosocial forms of embedding. We contend, however, that moving imagery is important precisely because it is endowed with special properties in relation to authenticity, anticipation, persuasion and disciplining.[3] In doing so we draw on and contribute to existing scholarship that has focused on the dissonances of emotion management in film (for example, in relation to fighting apathy, sustaining attention and avoiding panic).

Here we draw inspiration from Carey's (2008, 1989]) idea of a balance between two views of communication: the ritual view and the transmission view. The films we study cannot be reduced to containers of information about a dominant imaginary. Embedding through moving imagery may (on the ritual view) be associated with and reflective of the continuous creation of community, fellowship and social norms. Repetition is central to such practices. The transmission view, on the other hand, is more closely tied to instruction, often for purposes of disciplining and control. Although both views of communication are typically present, the analytical value of the distinction lies in the recognition that 'communication is a symbolic process whereby reality is produced, maintained, repaired, and transformed' (Carey 2008: 19). This is no less true of visual communication. Hence, we pay attention to an entire life cycle of film production, from decision, financing, production and dissemination right up to the social and cultural practice of viewing. We focus especially on the latter, identifying a gradual shift during the 1950s away from films circulated through civil defence organisations and cinemas towards the comparatively new mass medium of television. We identify an overlap between a shift of content and message and a shift in media and circulation. The transition away from ritualistic methods of organised public film screenings towards transmission of information into the home of citizens was accompanied by a new balance of responsibility for in civil defence. In short, our point is related to McLuhan's (1964) famous dictum that the medium is the message.

Our second and third contributions require less elaboration. Although Jasanoff (2015b: 226–230) points out that embedding is not a smooth process, the place of this phase alongside other phases in the framework introduces a risk of flattening dissonance and crises. By reading films in the specific national contexts in which imaginaries of civil defence were conceived, practised and promoted and by focusing on a period of contestation and transformation, we highlight crises and complexity. Moreover, our attention to the connection between media and messaging can shed light on (some of) the ways in which imaginaries undergo transformation. Given the severity of the challenge to the concept of civil defence during the 1950s, we examine *how* and *how successfully* moving imagery sought to assist the imagination of audiences in negotiating leaps between the present and the future, fear and hope, realism and utopianism. Our focus in this endeavour is predominantly on the content of the films, yet we also pay attention to pace, editing and music where relevant.[4]

Finally, we draw on Lee Clarke's concept of 'fantasy documents', which was originally developed with reference to detailed plans for managing disasters. One of the properties Clarke ascribes to fantasies is an ability to shape meaning and behaviour, a perspective that is compatible with the concept of imaginaries. Clarke's analysis, however, emphasises the role of assumptions and scenarios. Indeed, fantasy documents seek 'to transform uncertainties into risks, rationalizing problems previously outside the realm of systematic control' (Clarke 1999: 11). Some civil defence films produced in the 1950s and early 1960s can be read as fantasy documents, we suggest, exactly because they presented 'live' crises to which citizens and organisations responded (and, in the case of the films we have studied, successfully so). The most potent method by which the films achieved this effect was by specifying threats through scenarios. These anticipations of the crisis were typically informed by assumptions that, although within the bounds of possibility, were both contestable and also, especially in the early 1960s, contested. Approaching filmmaking from this perspective brings out more clearly its role in (successful or unsuccessful) embedding practices.

It is with these theoretical ambitions in mind that we approach Danish and Swedish films produced or co-produced by civil defence authorities or volunteer organisations (working closely with the state). The number of Danish films—produced by the Danish Civil Defence Directorate (DCDD, *Civilforsvarsstyrelsen*) and the Danish Civil Defence League (DCDL, *Civilforsvars-Forbundet*) with state involvement[5]—is modest, and we have included most of the 10–12 films produced during the period. The number of Swedish film productions is considerably higher (+30) and harder to specify, since several of these were co-productions between a wider range of authorities (including the military and the fire-fighting services). Not all of these films, however, were intended for the wider public. Following a screening of a majority of the available films and a process of triangulation with civil defence journals, newsletters and archival material, we identified a number of films that we have used as the primary materials for our analysis.[6]

Early films (c. 1949–1954)

Both Swedish and Danish civil defence authorities placed a premium on the use of film during the early 1950s, and they spent a considerable amount of energy and funds on producing films for purposes of recruitment and propaganda.[7] Danish authorities were under financial constraints, whereas the Swedish system was more flexible and better funded (Norén 2012: 35–46). In both countries, however, there was considerable continuity both in relation to the use of the film medium stretching back to the interwar period and in relation to an established framework of Second World War aerial defence. The films also display many of the traits that are associated with the early development of the genre (Aufderheide 2007: ch. 1). Dramatic sound (drums, shrieking bombs, air-raids, music), a voice of God and an instructional modality are central characteristics.

In productions from both countries, the specific threat posed by the atomic bomb plays only a marginal role. Despite discussions about the radical nature of the atomic bomb, within civil defence organisations it gradually came to be seen in the framework of the strategic bombing known from the Second World War. Radioactivity was often mentioned as a special property of this technology, but it was not accorded much weight. Such factors conspired to keep much planning and certainly a substantial part of the existing imaginary of the next (aerial) war intact. Hence, civil defence authorities incorporated the atomic threat into an existing set of practices and maintained the emphasis on collective organisation.[8] In other words, Danish and Swedish civil defence organisations functioned essentially in the same way in 1950 as they did during the war—only they were more extensive, better equipped and better drilled. War loomed large in the background, and it was unquestionably conceived to be catastrophic. But it remained an un(der)specified phenomenon, corresponding to a sense of unease and *Angst*, which recruitment efforts appealed to.

These features were mirrored in the films' manner of circulation. Both countries struggled with the dissemination of instructions and propaganda in written format and soon turned to other media (Bennesved and Norén 2020; Sylvest 2018). Moving imagery came to play an important role here, yet the specific circumstances of viewing are significant.[9] Screening films in social settings involved both limitations and a normative commitment that fitted the underlying imaginary. In local associations, film screenings were cumbersome, since they required expensive equipment and film distribution. To solve this problem, civil defence organisations in both Sweden and Denmark began offering cheap or free-of-charge rental services by mail as well as film tours.[10] Still, the function of films hinged on the willingness of citizens to gather and respond to the message, whether in local associations or at the cinema (where audiences were larger but presumably also less motivated). Local associations constituted an important site of screening in which recruitment was combined with curating, educational activities and exercises in a social setting in an effort to inspire and impose norms.[11] Strikingly, this way of organising and distributing early films reflected their core message: civil defence was seen as a collective effort in which it was incumbent on citizens to gather and organise under the watchful eye of the authorities in order to help themselves, their fellow citizens and their country during war.

Looking closely at the content of the films, civil defence is portrayed in part as an insurance policy against unforeseen events and in part as a natural extension of existing social practices. There are few references to the specific nature and causes of war, but the films imply that the *next* war will be total in a new way; yet this phenomenon is typically visualised by scenes from the *previous* war.[12] Consequently, a vague fear of the Cold War turning hot is pervasive, while the arrival of war is described as something beyond national control. The embedding efforts of these early films are dominated by an elaborate, state-led social organisation of human resources in protecting the homeland. The rationale and efficiency of collectively organised civil defence is displayed in a myriad of ways, but some are especially prominent and revealing of the imaginary informing these newly reorganised and renamed organisations. Four interrelated strategies stand out: the display of technologies and material resources; the seamless nature of the civil defence organisation and its transition from peace to war; a powerful image of efficiency based on the utilisation of technology and human resources; and, finally, a representation of volunteering as symbolic of a positive, imagined future for citizens.

The first of these strategies for communicating a message of competence and reassurance manifested itself in the display of a plethora of modern technologies, ranging from aircraft, trains, trucks and telephones to more mundane artefacts such as fire hoses. The distinctively modern look of specific artefacts or their mere abundant existence was used to signal organisational efficiency. In the Danish film *Protecting the Population* (*Befolkningens beskyttelse*, 1953), for example, almost a minute of the ten-minute information and recruitment film is taken up with panoramic views of the technologies necessary for facing the spectre of modern war. Swedish films communicate a similar message by zooming in on a civil defence organisation operating underground. Displaying hi-tech equipment and the operation of collective organisations was central to the imaginary of defending the homeland. An emphasis on scale hints at the national purpose and positions of the Danish and Swedish authorities (local and state) as guardians of its citizens. Identification is achieved through elaborate descriptions of the organisation of civil defence (including its various functions and branches, as well as charts of local organisations) and demonstrations of the smooth efficiency of the machinery in action. Among several objectives, the films sought to instil confidence in a collective, technocratic and hierarchically organised effort.

The second strategy revolves around the seamless nature of the collective social organisation and its transition from peace to war. This logic is visualised in the Swedish film *Bombed Town* (*Bombad stad*, 1949), a popular film at the time.[13] A highly technologised set of activities are used to demonstrate the efficiency of civil defence, from military radar observation centres and alarms to the evacuation of citizens into urban air-raid shelters and firefighting activities in the aftermath of an air-raid. Reflecting a general trend, the scenario is based on a conventional (rather than atomic) attack from the air ('a massive attack, around 50 planes' as it is stipulated in the Swedish production *Bombad stad* from 1949). Core themes of anticipation, control and order shape the imaginary. Phones, trains, observation posts and air-raid shelters are not only perfectly organised, but also function flawlessly in concert with personnel that under strict emotional control handle everything by the book. In these highly stylised but underspecified scenarios, the early films acquire some of the qualities that Clarke associates with fantasy documents. Thus, when the bombs fall, the element of surprise is sometimes missing—both for the enemy and the audience (in one Swedish production, the evacuation train even leaves shortly before the alarm goes off). A ticking clock in the background emphasises order and anticipation, adding to the image of control. Consequently, the damage is reduced, lives are saved, and the necessity and efficiency of civil defence are validated.[14]

The first and second strategies are both undergirded by a strong social (and ethical) creed that spoke to the national communities and the bonds between citizen and state, typically through simple yet powerful analogies between the private and the public spheres. *Protecting the Population* (DEN 1953) explicitly extended care in the private sphere to the entire national community. Having opened with displays of maternal care and vigilance in relation to children, the film ends by admonishing the audience that ethics of order, vigilance, preparedness and humanitarianism 'is so natural when it concerns the little ones, but it also applies to grown-ups, for everyone, for all in this country that is worth preserving'.[15] Similarly, in the Swedish film *Evacuation of Urban Centres* (*Utrymning av tätort*, 1950), evacuation routines clearly mimicked conventional forms of social organisation. Families were split from individual family level up to the societal level; men were sent into the military organisation and women into the social and healthcare system, while children were evacuated from the cities with their schools. In a similar manner as Marie Cronqvist identified a civil defence organisation of the 1950s saturated with welfare-state symbolism, an imaginary is promoted and sustained in these films through existing social norms and moral coda.[16]

The third strategy on display in the early films revolved around a tight bond between social organisation and the technologies that saturated the imaginary. Human resources are not only technologised, but are also compartmentalised so that they can be harnessed within a hierarchical, specialised structure. As a result, the efficiency of civil defence organisations is amplified. Humans are equipped with and master modern technologies of transport, communication and firefighting. Officers use phones for orders and reports, and workers in uniforms use drills and welding equipment to free trapped civilians. A scene in the Swedish film *Bombed Town* subtly connects the social and technological dimensions of civil defence. In a meeting between an elderly couple, the Berggrens, who had their homes bombed, and a young female civil defence volunteer in a civil defence service centre, Mr Berggren's social status is described with reference to his technical skills. Consoling the bereft couple, the young woman smilingly exclaims that Mr Berggren is a welder, and thus worth his 'weight in gold'. The couple remains uneasy but nevertheless accepts their situation while the young civil defence official ends the conversation by shouting: 'NEXT!'[17] The conversation here not only valourises the technical and technological skills of workers, but also highlights how the (gendered) social skills of the young official allow her to console fellow citizens, contain their discontent and inspire a restrained, positive attitude.

Fourth and finally, since public confidence in civil defence and/or the recruitment of volunteers were constant concerns, early films celebrated the humanitarian spirit of volunteers and their devotion to the noble cause. In depicting imagined futures where such qualities were on display, filmmakers walked a tightrope. They played on fears of war and catastrophe but also used this nightmare vision for socially useful, regenerative purposes. This Janus-faced character of embedding the imaginary was apparent from the outset, and it had implications for recruitment strategies. In the Danish case, the initial plan was to recruit hundreds of thousands of volunteers.[18] The very first civil defence film sought to draw them in by playing on fears of war and using 'powerful effects' while maintaining what the director called a 'positive' tone.[19] In several productions, the narrative revolved around a protagonist of civil defence work encountering an apathetic or ignorant character in need of conversion.[20] The films provided that by pressing the analogy between good and rational behaviour in everyday life and in circumstances of catastrophic war. In the Swedish case, in which recruitment was less of a problem because of compulsory drafting from 1944, citizen confidence in state efforts was of greater importance. In the closing scene of *Bombed Town* (1949), a couple going to bed reflects on their anger with compulsory civil defence activities the previous year. Having survived a bombing raid, however, they express their gratitude towards the state. In *Bombed Home* (SWE, *Bombat Hem*, 1950) viewers are presented with a tense and slightly confused atmosphere in a shelter during an (unspecified) aerial attack. Order is imposed by a uniformed woman who directs the sheltered citizens and instructs fellow civil defence personnel.[21] In such contexts, the uniform functioned as a symbolic and material manifestation of successful recruitment and as a marker of competence and authority. Displays of authority, however, were also tempered by the commonality among citizens and civil defence personnel. Towards the end of the film, uniformed and non-uniformed citizens emerged arm-in-arm from the ordeal into a destroyed urban area.[22] Such messages of reassurance achieved through a display of both formal, skilled authority and community and familiarity were common in early filmmaking.

In sum, the core message of early Swedish and Danish films revolved around the mantra, ritually repeated on screen, that collectively organised preparation for catastrophe was essential, necessary, rational and the best available measure for avoiding panic and death. By mirroring the social order of society within a hi-tech civil defence organisation that valued technical skills and appealed to an existing moral economy in efforts to secure public support and draw in volunteers, these films sought to embed an imaginary of civil defence that displayed considerable continuity with recent experiences of war. The collective, uniformed and state-controlled national effort was deemed crucial for survival. Distinct roles were performed by individuals according to age, gender, class and function, yet this stratification served to reinforce a collective, national and humanitarian effort.

Later films (c. 1954–1965)

By the early 1960s, civil defence films in Denmark and Sweden had taken on a new character. They struck a different tone. The opening scene of the Danish production *Radioactive Fallout* (DEN, *Radioaktivt nedfald*, 1961) is an instructive example. Set on a beach to unnerving music, the speaker addresses the viewers directly and asks them to reflect on the possibility that what we fear most—the dropping of an atomic bomb—may become reality. 'You don't want to imagine this? Is that what you are thinking?' The film then proceeds to assist the imagination of the viewer. The transformation embodied in *Radioactive Fallout* did not occur overnight, and it arguably had its roots in the gradual and uneven response to the thermonuclear disruption of civil defence that was inaugurated by the US *Castle Bravo* test on Bikini atoll on 1 March 1954. Step by step, information and a string of worries about the H bomb made their way to the public sphere. In both Sweden and Denmark, it was soon apparent that this new weapons technology could potentially have enormous implications for civil defence, partly because of the sheer power of thermonuclear weapons and partly because of a new property that followed in its wake: radioactive fallout. Danish and Swedish authorities did not react immediately. The consequences of the disruption took time to gestate and only began to show towards the end of the decade following an intense period of information gathering and strategic deliberation. On film, it translated into a shift in tone and message: the later films focus explicitly on the nuclear threat, they struck a new balance between the responsibility placed on organised civil defence and that of the ordinary citizen, and they evinced a far more complicated and darker vision of life during and after a catastrophe.

Film kept its status as an important tool in embedding practices, as the rate of production testifies. Towards the late 1950s, the number of productions went up, and increasingly they moved away from the ideal–typical features of the early films. In Sweden, several productions dealt with the exposure of urban environments to atomic weaponry, evacuation and the Swedish air-raid shelter system (especially large-scale population shelters and radioactivity).[23] Danish productions, though smaller in number, display a similar pattern: films produced during the late 1950s and early 1960s were more forthright in acknowledging the nuclear dimension of civil defence.[24] The changing message was accompanied by a shift in production and screening practices in both countries. Films were still made for cinemas or available for screening in local associations, but civil defence organisations increasingly turned towards the television medium. In 1958, Swedish authorities also rejected a series of films produced during the early 1950s, branding them outdated and inappropriate.[25] In Denmark, the transition was more gradual, and productions up until 1959 continued to pass on advice of the *Duck-and-Cover* variety that US authorities had declared retrograde.[26]

This shift in the dominant imaginary of civil defence was mirrored by changes in the media landscape and the prevalent view of communication. In contrast to the early period, where the circulation of civil defence films was marked by a ritual form of communication, the hydrogen-induced age of risk management eventually became characterised by a trend towards communication as transmission. The turn to television in communicating preparedness to wide audiences signifies the shift. Eager to find new ways of disseminating authoritative information, authorities began educating and instructing all citizens, enabling them to secure their own survival. Famously, this was done through pamphlets distributed in several Western countries during the early 1960s (including Germany and the UK), but television broadcasts also became increasingly popular. Moreover, these strategies also worked in tandem. State television functioned as visual companions to the mass distributed leaflets. In Denmark, the pattern was similar, although as a small country it placed less emphasis on mass evacuation.[27]

Where previously films had sought to draw citizens into a social organisation that would provide detailed information (and exercise) only in a second step, specific advice was now funnelled directly into the homes of ordinary citizens. Local civil defence organisations no longer functioned as mediators. Consequently, a symbolic gap emerged between the safely bunkered civil defence broadcaster and the public recipient. Recruitment was no longer a top priority. Authorities would do their utmost to assist the citizen in the struggle for survival, but the buck stopped with the responsible citizen. Efforts to embed the imaginary were, therefore, aimed at the 'general public'—typically understood as a household father and mother facing a television set in the relative comforts of their home.

The new political and technological context in which civil defence operated was also clearly reflected on screen. Gradually, from the mid-1950s, the *raison d'être* of civil defence became a matter of (some) public controversy. Swedish civil defence witnessed a major change in tactics that reflected the significance of the hydrogen bomb and new means of delivery that drastically reduced warning time. At the same time, debates about the nature and properties of radioactive fallout transformed civil protection. During a period when individual responsibility attracted more attention, collective civil defence in Sweden came to revolve around two distinct methods of protection: preparedness evacuation of whole urban environments and large-scale nuclear-proof population shelters for those who remained (Cronqvist 2012, 2015; also Bennesved and Norén 2020). The effectiveness of both measures was, however, in dispute.[28] The shift in Danish policy was slower and less explicit but was also reflected in public debate (Sylvest 2020). By the late 1950s, some peace activists and radicals viewed civil defence as wildly utopian in its focus on local, atomic (rather than thermonuclear) warfare, and even when civil defence organisations explicitly addressed questions of fallout, the reception was mixed. While several newspaper reviews, for example, commended *Radioactive Fallout* for providing specific advice, critical voices—from both left and right—questioned the film's scenario and implied that civil defence was outmoded or utopian.[29] Thus, public debate in both countries increased the pressure for information and solutions, and authorities were increasingly criticised from two directions: one flank accusing civil defence of being in league with militarism and the other accusing it of being inadequate and outmoded.

This burgeoning criticism shaped the focus, message and format of films produced in the late 1950s and early 1960s. It is reflected in three features of filmmaking in the period. First, the films produced during this period often explicitly focused on the threat posed by atomic or nuclear weaponry. Two ideal-typical forms of this shift of focus are apparent (some films, of course, displayed a mixture of these). On the one hand, a number of films focused on the role of civil defence in the event of a smaller atomic attack, a scenario that displayed continuity with the portrayal of civil defence as a modernised air-defence structure. In such films, however, it was increasingly made clear that the speed of modern warfare made it essential for every citizen to ensure their own immediate survival—often through simply acquiring information on where to go. In Denmark the production *Civil Defence* (1959) is an example focusing on the general population's immediate survival through air-raid shelters and evacuation.[30] In Sweden, films such as *Also a City Problem* (1960) similarly emphasised the necessity of having 'simple instructions' for the 'individual' in situations in which speed was of the essence.[31] On the other hand, productions of the early 1960s tended to focus on the more complex scenario of defending Denmark or Sweden against the effects of thermonuclear war. Here war—rarely if ever involving a direct H bomb attack on either country—was portrayed as a meteorological phenomenon, with radioactive fallout silently and invisibly striking the country (rather than being the direct effect of an H bomb explosion). The most iconic films in this respect are *Radiak* (SWE 1962) and *Radioactive Fallout* (DEN 1961). These films are at once more futuristic, vague and sombre, and they involve an element of public education for the nuclear age.

Second, it was during this period that civil defence films evinced a tendency to individualise responsibility for survival, which in turn transformed the role(s) of collective civil defence. The stress on individual responsibility had been apparent for some time in the USA, but with the hydrogen bomb, it gained ground across countries in the West. Significantly, NATO committees concerned with civil protection began pushing this agenda in the mid-1950s.[32] Recruitment into volunteer organisations became less important. A redistribution of responsibility—generally speaking, from collective towards individual responsibility—involved difficulties. These problems concerned science education (necessary for understanding the properties of nuclear weapons technology) and how to make the effects of the H bomb intelligible. In these matters, filmmaking had to chart a course between inducing action and avoiding narratives that risked undermining the imaginary. Hence, it became a common strategy to use abstract visualisations, maps or other forms of modelling and experimental imagery. The properties and dangers of radioactivity and fallout were often visualised through drawings and cartoons.[33] Similarly, the nature and effects of thermonuclear weapons were typically demonstrated by animations of nuclear explosions, sometimes in combination with footage from (US) tests. Despite their abstract nature, such sequences could be chilling.

Third, it was a characteristic strategy of the later films in both countries that they sought to mitigate the horrors of nuclear war by obfuscating politics and the causes of war while installing an ever-present risk of nuclear war into orderly and ordinary everyday life. Clearly, new weapons technologies placed new demands on civil defence, but the arrival of (thermo) nuclear war (just like the fallout that would follow) appeared as wind, clouds and rain with temporary effects. So while the risk of war is a pervasive feature of these films, the political nature of war is oddly absent and disconnected from the main message that preparedness is (also) an individual responsibility. Taken together, the films are emblematic of a drift of policy in Swedish and Danish civil defence at the time. In Sweden, self-protection was launched in earnest after 1960–1961.[34] Soon after, in early 1962, Danish authorities finally informed the public about the dangers of (thermo) nuclear weapons and the consequences of their use. The filmic strategies designed for embedding this incarnation of 'preparedness' revolved around calm, well-prepared and methodical citizens taking responsibility for themselves and their families in times of crisis.

When every citizen became a target—also of information—their conduct was depicted in new ways. In one Danish film from the period, the speaker intones that 'in modern war, the warning will be so short that you must seek shelter immediately'.[35] The housewife, the farmer and the worker are typically unfazed on receipt of warnings about incoming attacks or fallout. Attentive to the instruction of authorities, they go about their duties in a highly disciplined fashion, whether they concern the safety of children, livestock or co-workers. In Sweden, the household car became a key technology that would ensure survival if kept with a full tank. This was, again, an individual responsibility decoupled from organised civil defence services. Television shows such as *If the War Comes* (SWE, *Om kriget kommer*, 1961) emphasised the housewife's role in processing instructions into practical pre-packed suitcases and evacuation routines.[36] At some points, the instructions communicated through these examples produced dissonance and paradox. Indeed, some strategies of survival appeared to be remnants of the 'atomic age', if not outright inadequate, in the face of thermonuclear war: run, hide, close the door, wash yourself, and clean your cupboard. Yet, they could be justified within the parameters of official scenario planning. Still, the overall point—that citizens were to a large extent responsible for their own survival—appealed directly to social norms about responsibility in the family and the workplace. This is unlikely to have been lost on audiences.

Similarly, when every citizen became a civil defender, the collective dimension of civil defence took on a new guise. On screen, uniformed personnel still played a crucial role in defending the homeland and the population, but they increasingly appeared as a (bunkered) elite of cool experts that provided an early warning, directed or assisted evacuation efforts (if time permitted) or monitored levels of radioactivity. The siren was central and used to dramatic effect as both audio background and as a visual symbol of danger. Many sequences are rich with artefacts of organised civil defence; uniforms, trucks, firefighting equipment and communications technology are amply on display. Tactical considerations and decisions took on a prominent role; radar detection of enemy airplanes and telephone conversations appeared as central nodes in a network of activity that would ensure survival. Civil defence was still emblematic of the well-prepared society. In contrast to the earlier films, however, we now encounter visible signs of strain and exhaustion. As scenarios of catastrophe unfolded and as they were extended into an imagined future, life underground was depicted as manageable but also taxing and bleak.[37] So while the distance between civil defence authorities and citizens increased and the balance of responsibility was reconfigured, the later films included an admission of the enormity of the challenge and its likely impact on both citizens and authorities. This increasing 'realism', however, was combined with a dogged insistence on the rationale of civil defence in the event of atomic and thermonuclear war. The significant leap between these two components of the imaginary was overcome by the use of highly specific scenarios. In conjuring up an entirely new form of warfare, assumptions were made about the nature and yield of weapons, the target, the weather, the resulting contamination and, even, the nature of the unforeseen circumstances that would occur.

It was at this point especially that these later productions—viewed as embedding efforts—came most closely to resemble fantasy documents. In a context of public scepticism and (assumed) apathy and faced with a new, formidable military technology, the problem was rationalised and brought under control through scenario building and simulation, often in specific, localised settings. This was not an entirely new strategy. Earlier films also operated with idealistic assumptions about an unfolding war, and there was still a premium on control in demonstrating efficiency and the necessity of the system. Crucially, however, the war depicted in the earlier films was a war that audiences understood through experience and collective memory. In contrast, the nature of thermonuclear war lay outside the realm of experience even as the fear of it could be very real. This was a premise of the later films that was reflected in their darker tone: personnel in underground structures, for example, continue to perform their duties despite the eradication of their entire hometown and its surroundings.[38] Here, we encounter another dissonance: the later films are in a sense less fictional and achieve a quasi-realist effect by zooming in on hardship, unpredictability and the hazard of invisible, radioactive contamination, yet this effect was a product of an elaborate use of rationalisation through scenarios that, eventually, insisted on the possibility of survival. In the process, pressing questions about the nature of life in a postwar world were ignored. It was as if the crisis was over once measurements of radioactivity had returned to safe levels. In that sense, the films are not merely fantasy documents but also fantasy monuments.

Conclusion

Our study has shown that Swedish and Danish filmmaking during 1949–1965 reflected a transition in both the imagination of future warfare and the rationale of civil defence. Following the thermonuclear disruption of the mid-1950s, the central imaginary of civil defence in both Sweden and Denmark underwent significant change in an effort to retain credibility. Although several mantras of civil defence persisted—including a dogged insistence on the value of preparedness and preparation—during the late 1950s and early 1960s, the national, collectively oriented legacy of aerial defence was transformed in both countries. A shift towards individualisation can effectively be read through film and television, which constitutes a rich source for studying imaginaries of civil defence and their embedding. Films produced towards the end of the period placed more emphasis on instruction and detailed information, while seeking to embed this new set of responsibilities in modernising societies and through shifting media cultures. They did so at least in part as a response to mounting political pressure for more information arising from national, transnational and international anti-nuclear movements.

The theoretical approach lodged in the concept of sociotechnical imaginaries has been productive especially in attuning us to the nature and dynamics of films as a form of psychosocial embedding. Moreover, it has allowed us to bring out just how closely technology and social norms were imbricated in visions of the next war. Still, important analytical and theoretical questions remain, and we are not absolved of grappling with thorny issues about the demarcation of imaginaries, the inescapable overlap between embedding and extension, or the precise influence, reception or effect of a particular imaginary. It is very hard to say something definite about such matters, and most empirical studies will be faced with questions about the representativeness of the source material and the reach of conclusions. In the context of empirical studies, these kinds of theoretical debates will persist and are to be welcomed. Yet, there are also more glaring pitfalls to avoid. In particular, the risk of producing a teleological narrative that reduces change to technological 'progress' appears especially pertinent. Staying true to the underlying logic of co-production in the analysis of sociotechnical imaginaries is demanding, not least because it is a common inclination to identify social effects of specific technologies. The reverse dynamic—how social practices, norms, identities and institutions shape specific understandings of or responses to a technology—appears less habitual.

In our analysis, the process of (visual) individualisation of responsibility and its resulting changes in imaginaries of civil defence were not simply epiphenomena of technological change. Indeed, civil defence communication also shaped the public imagination of nuclear war and survival. The extent to which the films influenced the behaviour of the 'ordinary citizen' is impossible to determine, yet we should perhaps be wary of thinking too much along such lines that sustain a positivist approach to (visual) propaganda, information campaigns, or really any form of public narration by state representatives. In contrast, it may be more productive to approach these films as *attempts* at embedding. We are reasonably confident about the contours of the imaginary that civil defence authorities sought to advance, but less so about its subsequent fate. Here, we may lack a critical edge. To the extent that the approach and especially the sequential logic of the four phases in the life cycles of imaginaries appear infused with a logic of direction, progress, and perhaps even success, we ought to question it.

Indeed, the late 1950s and early 1960s can be read as the first phase of a long period of crisis that coincided with a growing criticism in Western Europe (and the USA) towards the self-proclaimed omnipotence of tradition and authority vested in state apparatuses. From this perspective, visualisations of the dominant imaginary of civil defence were rife with dissonances that, gradually, undermined its credibility. Individualisation was not a set trajectory, but perhaps rather a slow *withdrawal* from responsibility that pointed to a deeper crisis of faith inside and outside the ranks of civil defence organisations. Public criticism, declining membership numbers (Sweden had 660,000 volunteer members in 1945, and only 10,000 members in 1989[39]), decreasing state funding, rhetorical retreat in the face of Mutually Assured Destruction, and an orientation towards new peacetime activities point not to a mature and institutionalised imaginary of civil defence but rather to a lacklustre organisation of all-hazards risk management that could be openly defied and ridiculed. When civil defence organisations faced public scrutiny again in the 1980s, they certainly encountered such attitudes. While the emphasis on technology and the future that feature so prominently in the concept of sociotechnical imaginaries is clearly productive, there is also a need here to disturb 'the tidy timelines of progress' (Edgerton 2006: xii) that are associated with technology and stories of invention and adjustment. We are likely to gain more analytical power by approaching imaginaries of civil defence not as cyclical processes of successful adjustments and refinement but as a combination of ingrained reflexes and controversial political imperatives. From this angle, the imaginary of civil defence repeatedly encountered dissonance and crisis when it created images that reconciled life in the modern world with the dreaded future of nuclear war.

Notes

1. Sweden had already merged all civil defence activities into one organisation in 1944, whereas Denmark established its civil defence by law in 1949.
2. See also Bjørnsson (ch. 2) in this volume. See also Oakes and Grossman (1992: 394–395) and Collier and Lakoff's discussion on the birth and evolution of 'Emergency federalism' (2008: 13–14). In relation to Sweden, Cronqvist (2012: 205–206) makes the case that civil defence did not experience 'privatisation', but perhaps 'state individualisation'.
3. We leave aside the difficulties of separating the dynamics of embedding and extension that emerge from Jasanoff's discussion (2015b). We appreciate the attempt to specify these dynamics, though they are often intertwined in (historical) practice.
4. Historians must pay attention to both aspects, or what Platinga (2005) terms the 'saying' and 'showing' parts of documentaries.
5. The relevant state agencies included The Film Committee of Danish Ministries (*Ministeriernes Filmudvalg*), Danish Cultural Films (*Dansk Kulturfilm*) and The Danish Government Film Office (*Statens Filmcentral*).
6. Swedish films are available through the Swedish Media Database, www.smdb.se [accessed 1 April 2020]. Most Danish productions are available through the Danish Film Institute: https://www.dfi.dk/en/english/danish-film-history [accessed 1 April 2020].
7. For Denmark, see DCDL to The Film Committee of Danish Ministries, 15 December 1950, in Old film records, 'Civilforsvaret', The Danish Government Film Office, Danish National Archives (DNA)/The Danish Film Institute (DFI); for Sweden, see Norén (2012: 43–45).

8. See Bennesved (2020). This 'conventionalisation' of the atomic bomb and its reception into an existing set of practices comes out in several chapters in this volume. See especially Bjørnsson (ch. 2) and Farbøl (ch. 7).
9. Danish films were specifically produced for either cinema distribution or local associations. DCDD, *Status for civilforsvaret*, September 1953, 27–28. DNA, Schultz Memoranda 185. For the film plans of Swedish civil defence, see Norén (2012).
10. Indeed, from the mid-1950s, the annual reports of the DCDL contained information about the most popular films, and film catalogues were printed to service local associations.
11. See, for example, the following Swedish report from 1953: 'Civilförsvarsupplysning genom utomhusbio', *Tidskrift för Svenskt civilförsvar* (1953), 75–76. For an analysis of similar trends in the USA and the UK, see Matthews (2012) and Hogg (ch. 4) in this volume.
12. See also the chapters by Bjørnsson (ch. 2) and Hogg (ch. 4) in this volume.
13. 'Från arbetsfältet', *Tidskrift för Svenskt civilförsvar* (1950), 22; See also letter to Arthur Dahl 1 October 1949, Vol. B VII: 14 'Brev från Danmark' Civilförsvarsstyrelsens öppna arkiv, Krigsarkivet, Swedish National Archives, Stockholm, Sweden.
14. However, on overly political matters, like the risk of Sweden maintaining a neutral and alliance-free stance in the postwar period, *Bombed Town* is silent.
15. *Protecting the Population* (DEN 1953), 10:05–10:15.
16. For more on the connection between Swedish civil defence and welfare politics, see Cronqvist (2008, 2012).
17. *Bombed Town* (SWE 1949), 18:00–18:40.
18. In 1951, the goal was to make 600,000 Danes 'civil defence conscious' (DCDL to The Film Committee of Danish Ministries, 25/6–1951. Old film records, 'Mit navn er Christensen', *The Danish Government Film Office*, DNA/DFI).
19. 'Draft of propaganda film for Civil Defence', Old film records, 'Civilforsvaret', *The Danish Government Film Office*, DNA/DFI.
20. See for example *Women in Civil Defence* (DEN, *Kvinder i Civilforsvaret*, 1952) and in *My Name is Christensen* (DEN, *Mit navn er Christensen*, 1953).
21. *Bombed Home* (SWE 1950), 15:50–17:11.

22. *Bombed Home* (SWE 1950), 29:00–29:05.
23. See for example *We Move Underground* (*Vi går underjord*, 1959), *Also a City Problem* (*Också ett storstadsproblem*, 1960) and *Radiak* (1961). The CDA also produced a series of evacuation films aimed at audiences in dense urban areas.
24. See especially *Civil Defence* (*Civilforsvaret*, 1959), *Air Attack* (*Luftangreb*, 1959) and *Radioactive Fallout* (1961) – the three most significant films of the time.
25. 'Nytt från Civilförsvarsstyrelsen, Kasserade filmer', *Tidskrift för Svenskt Civilförsvar*, (1958), 26.
26. See especially *Air Attack* (1959). On US advice, see Titus (1983: 6).
27. In this period, civil defence authorities in both countries also began producing and participating in television debates. In Sweden, following the successful show *Our Need for Security* (*Vårt behov av trygghet*) in 1958, the CDA decided to work with television to launch a new self-protection campaign in 1961. Similarly, the previously mentioned film *Radiak* (1962) was from the start intended as a television broadcast: 'Radiak – tre nya filmer', *Tidskrift för Svenskt Civilförsvar* (1962), 85. *Radioactive Fallout*, the most ambitious and well-crafted Danish production, was broadcast on Danish Television on 6 December 1961, a month before the national distribution of the Danish leaflet.
28. In Sweden, criticism against civil defence was spearheaded by Aktion mot Svenskt Atomvapen, (AMSA), see Andersson (2001). Two veteran officers also criticised the then civil defence organisation as hopelessly inadequate causing much debate in public. See also Bennesved and Norén (2020).
29. See, for instance, reviews ('Rystende Realisme') in *Jyllands-Posten*, 8 December 1961 and ('Fra vort kælderdyb') in *Frit Danmark*, 1962, 20: 10, 14–15.
30. The film marked the 25-year anniversary of Danish civil defence. It followed an exercise based on a scenario in which a provincial town was attacked by a 'small atomic bomb'. Although there were clearly grounds for scepticism, the director sought to 'convince the viewer' of the effectiveness of civil defence in part through 'emotional appeal'. Carl Otto Petersen, 'Preface to civil defence film synopsis', 27 September 1958, Old film records, 'Civilforsvaret', *The Danish Government Film Office*, DNA/DFI.

The film contained a much-ridiculed sentence from a failed information campaign of 1958; namely 'that the best protection is naturally to not be where the damage occurs' (4:10–4:16).

31. *Also a City Problem*, 1960, 12:50–13:10.
32. See Bjørnsson (ch. 2) in this volume.
33. *Radioactive Fallout* (DEN 1961) was commended for its animated, technical explanation of radioactive fallout.
34. Self-protection ('Självskydd') was officially endorsed in 1960 by the Swedish civil defence services, in preference to the older 'Hemskydd' model.
35. *Civil Defence* (DEN 1959), 3:15–3:23.
36. Broadcast on Swedish television on 10 October 1961.
37. See, for example, *Radioactive Fallout* and *Evacuation of Västerås* (SWE, *Utrymning av Västerås*, 1960).
38. For example, *Our Need for Security* (SWE 1958), 12:00–16:00.
39. Hans C. Pettersson, in Sjölin (2014: 207).

References

Unpublished Sources

RA: Rigsarkivet (Danish National Archives)
1261: Beredskabsstyrelsen, Civilforsvarsdirektør E Schultz' embedsarkiv.
185: Memoranda
RA: Rigsarkivet (Danish National Archives)/The Danish Film Institute
Statens Filmcentral (The Danish Government Film Office)
Old Film Records
KRA: Krigsarkivet, Riksarkivet (Swedish National Archives)
0829: Civilförsvarsstyrelsen
Vol. B VII: 14
SMDB: Svensk Mediedatabas (Swedish Media Data Base): www.smdb.se

Websites

Danmark på Film (Denmark in Films): https://filmcentralen.dk/museum/danmark-paa-film
Filmarkivet (Sweden): http://www.filmarkivet.se

Literature

Andersson, Irene. 2001. En civilförsvarsvägran med rötter i 1930-talet. In *När Alving blev Bang*, ed. Marcos Cantera Carlomagno, 13–32. Lund: Historiska media.

Aufderheide, Patricia. 2007. *Documentary film: A very short introduction.* Oxford: Oxford University Press.

Baker, George W., and Leonard S. Cottrell, Jr. 1962. *Behavioral science and civil defense*, Disaster Study Number 16. Washington: National Academy of Sciences–National Research Council.

Bennesved, Peter. 2020. *Sheltered society: Civilian air raid shelters in Sweden – from idea to materiality, 1918–1940 and beyond.* Malmö: Universus Academic Press.

Bennesved, Peter, and Fredrik Norén. 2020. Urban catastrophe and sheltered salvation: The media system of Swedish civil defence, 1937–1960. *Media History* 26: 167–184.

Berger Ziauddin, Silvia, Peter Bennesved and Sarah Robey. 2018. 'Living (again) in an age of nuclear fear: New avenues for studying Cold War civil defense', 16 April 2018. http://www.berlinerkolleg.com/sites/default/files/robey_berger_ziauddin_bennesved_cold_war_civil_defense_en.pdf [accessed 21 January 2020].

Carey, James W. 2008 [1989]. *Communication as culture*, revised edn London: Routledge.

Clarke, Lee. 1999. *Mission improbable: Using fantasy documents to tame disaster.* Chicago: Chicago University Press.

Collier, Stephen, and Andrew Lakoff. 2008. Distributed preparedness: The spatial logic of domestic security in the United States. *Environment and Planning D: Society and Space* 26: 7–28.

Cronqvist, Marie. 2008. Vi går under jorden: Kalla kriget möter folkhemmet i svensk civilförsvarsfilm. In *Välfärdsbilder: svensk film utanför biografen*, eds Erik Hedling and Mats Jönsson, 166–181. Stockholm: Statens ljud- och bildarkiv.

Cronqvist, Marie. 2012. Survival in the welfare cocoon: The culture of civil defence in Cold War Sweden. In *Cold War Cultures: Perspectives on Eastern and Western societies*, eds Anette Vowinckel, Marcus M. Payk, and Thomas Lindenberger, 191–210. New York: Berghahn Books.

Cronqvist, Marie. 2015. Evacuation as welfare ritual: Cold War media and the Swedish culture of civil defence. In *Nordic Cold War cultures: Ideological promotion, public reception, and East–West interactions*, eds Valur Ingimundarson and Rosa Magnusdottir, 75–95. Helsinki: Aleksanteri Cold War Series.

Edgerton, David. 2006. *The shock of the old: Technology and global history since 1900.* London: Profile Books.

Grant, Matthew. 2013. Images of survival, stories of destruction: Nuclear war on British screens, 1945–c.1963. *Journal of British Cinema and Television* 10: 7–26.

Grossman, Andrew, and Guy Oakes. 1992. Managing nuclear terror: The genesis of American civil defense strategy. *International Journal of Politics, Culture and Society* 5: 361–403.

Jacobs, Bo. 2010. Atomic kids: *Duck and Cover* and *Atomic Alert* teach American children how to survive atomic attack, *Film & History* 40: 25–44.

Jasanoff, Sheila. 2015a. Future imperfect: Science, technology and the imaginations of modernity. In *Dreamscapes of modernity: Sociotechnical imaginaries and the fabrication of power*, eds Sheila Jasanoff and Sang-Huyn Kim, 1–33. Chicago: Chicago University Press.

Jasanoff, Sheila. 2015b. Imagined and invented worlds. In *Dreamscapes of modernity: Sociotechnical imaginaries and the fabrication of power*, eds Sheila Jasanoff and Sang-Huyn Kim, 321–341. Chicago: Chicago University Press.

Jasanoff, Sheila, and Sang-Huyn Kim, eds. 2015. *Dreamscapes of modernity: Sociotechnical imaginaries and the fabrication of power*. Chicago: Chicago University Press.

Masco, Joseph. 2008. Target audience. *Bulletin of the Atomic Scientists* 64(3): 22–32, 45.

Matthews Jr., Melvin E. 2012. *Duck and cover: Civil defence images in film and television from the Cold War to 9/11.* Jefferson, NC: MacFarland.

McLuhan, Marshall. 1964. *Understanding media: The extensions of man*. New York: McGraw-Hill.

Munster, Rens van, and Casper Sylvest. 2016. *Nuclear realism: Global political thought during the thermonuclear revolution*. Abingdon: Routledge.

Norén, Fredrik. 2012. Civilförsvarsstyrelsens filmaktiviteter under kallakrigsåren 1949–1952. *Historisk Tidsskrift* 132: 31–54.

Nowak, Lars. 2016. Images of nuclear war in US government films from the early Cold War. In *Understanding the imaginary war: Culture, thought and nuclear conflict, 1945–90*, eds Matthew Grant and Benjamin Ziemann, 260–286. Manchester: Manchester University Press.

Platinga, Carl. 2005. What a documentary is, after all. *The Journal of Aesthetics and Art Criticism* 63: 105–117.

Sjölin, Vilhelm. 2014. *I skuggan av kriget: svenskt civilförsvar 1937–1996*. Stockholm: Instant Book.

Sylvest, Casper. 2018, Atomfrygten og civilforsvaret, *temp – tidsskrift for historie* 16: 16–39,

Sylvest, Casper. 2020. Nuclear fallout as risk: Denmark and the thermonuclear revolution. In *Histories of knowledge in postwar Scandinavia: Actors, arenas and aspirations*, eds Johan Östling, Niklas Olsen and David Heidenblad, 21–38. London: Routledge.

Titus, A. Constandina. 1983. Back to ground zero: Old footage through new lenses. *Journal of Popular Film and Television* 11: 3–11.
Weart, Spencer R. 1987. History of American attitudes to civil defense. In *Civil defense: A choice of disasters*, eds John Dowling and Evans M. Harrell, 11–32. New York: American Institute of Physics.

CHAPTER 6

'The World is Her Home': The Role of Women Volunteers in Dutch Civil Defence in the 1950s and 1960s

Dick van Lente

The Dutch civil defence organisation Bescherming Bevolking (Protection of the Population), or BB, was launched, with great reluctance, by the Dutch government in 1952.[1] The first postwar governments estimated that the Soviet Union had been damaged too much during the Second World War to risk an attack on Western Europe. Communism was seen as mainly a domestic threat, which had to be countered by investing in a strong economy and comfortable living conditions for the whole population rather than wasting money on weapons. The Czech coup of 1948, the Berlin crisis of 1948–1949, the announcement in September 1949 that the Soviets had a nuclear weapon and the Korean War in 1950 changed the situation. The American government expected Soviet aggression in Western Europe, sent over more troops and demanded that the Europeans increase their military budgets. The Dutch government

D. van Lente (✉)
Department of History, Eramus University, Rotterdam, The Netherlands
e-mail: vanlente@eshcc.eur.nl

M. Cronqvist et al. (eds.), *Cold War Civil Defence in Western Europe*,
https://doi.org/10.1007/978-3-030-84281-9_6

complied. The new civil defence organisation was considered part of that defensive effort: it had to demonstrate the population's readiness to face an attack from the East. That, at least, was the theory. In practice, the BB elicited very little enthusiasm among politicians, officials and the wider public.

Like civil defence organisations elsewhere, the BB relied largely on volunteers, in addition to a small professional staff. Recruiting efforts were directed mainly at men, who would be trained in firefighting and rescue operations, while women were welcomed in the Division for Care and Comfort, which would help victims. As the number of volunteers remained far below expectations, the recruiting campaign was extended in 1955. The slogan of this campaign, 'the third man', testified to the male orientation of the BB. It indicated that one-third of the vacancies still had to be filled, but also, and more strikingly, that the organisation had simply by-passed the female half of the population.

When recruitments remained low, the BB realised its mistake, and in 1956, it began to target Dutch women. For this purpose, it sought the help of Nederlandse Federatie voor Vrouwelijke Vrijwillige Hulpverlening (the Dutch Federation of Women's Voluntary Organisations, FVVH). The FVVH agreed to organise courses on individual self-protection, but insisted on doing this independently of, although in coordination with, the BB. It quickly recruited and trained a large group of women instructors, and in the next decade organised hundreds of instruction meetings throughout the country. In the meantime, the BB languished: it had difficulty recruiting volunteers and keeping them involved. In the late 1960s, it practically faded from public view. This steady decline contrasted markedly with the gusto with which women volunteers undertook their job, and their relative success.

This contrast between a stumbling men's organisation and a thriving women's organisation, both working to prepare the Dutch population for nuclear war, is the topic of this chapter. Why did the women succeed where the men failed? The answer to this question, this chapter contends, is to be found in the way the women volunteers understood their role in society, and how they put this into practice. They were at the forefront of a feminist wave that rose after the Second World War, a movement that was situated between the more well-known first and second feminist waves of the fin-de-siècle and the late 1960s, respectively (Schwegman and Withuis 1993: 570). They were inspired by the early feminists, who considered the right to vote, acquired in 1919, as their entrance ticket to public debate, which would finally allow women to contribute to the improvement of society. The new realities of warfare, which since the First World War included aerial bombardment, gave women a new and urgent task. Military violence would no longer be limited to the frontlines, where men did the fighting, but would include homes, women and children. Therefore, in addition to feeding and clothing their families, women now had to protect them, a role which could extend from creating improvised air-raid shelters to getting involved in the peace movement and international politics. During the Second World War, many women played a crucial role in protecting and feeding their families, as well as providing shelter to people hiding from the occupying forces. This created a sense of collective strength and pride, and the conviction that they had an important role to play in postwar Dutch society. But postwar society turned out to be at least as patriarchal as it had been before the war, and the BB was an eminent example of a male-dominated institution. While needing the help of the women volunteers, the BB resented their independence and success. This conflict between the official imaginary of civil defence, incorporated in the patriarchal BB and the women volunteers enacting a new vision of the role of women in the nuclear age is the topic of this chapter (Jasanoff 2015a: 4).

The next section sketches the rise and decline of the BB. Section three discusses the debate about the role of women in society during the 1950s and 1960s. These two sections set the stage for the story of the voluntary organisations' efforts to teach women how to deal with the effects of nuclear war. The final section will evaluate to what extent and in which ways the women's voluntary organisations succeeded in enacting a new imaginary of gender roles for the nuclear age, and how these activities were received.

Rise and Decline of the Bescherming Bevolking

The Dutch civil defence organisation Bescherming Bevolking was created during a period of widespread fear of nuclear war. Anxiety levels rose after crises such as the Prague coup (February 1948) and the Berlin blockade (1948–1949), then lessened during the 1950s (Table 6.1).

The polls did not ask about the fear of *nuclear* war and its consequences, but judging by contemporaneous articles in newspapers and popular magazines, this was clearly implied. A popular illustrated magazine, *Panorama*, wrote in November 1946 that the next war would likely

Table 6.1 Polling on the perceived likelihood of war

Percentage of people polled who say yes to the questions below

	Will there be another world war?	*Will there be another world war within ten years?*
1945 December	50	32
1946 December	63	39
1948 September	71	52
1949 June	55	27
1950 September	60	41
1953 June (?)	31	11
1957 Sept	40	18
1962 15 Oct	43	20
1962 29 Oct	37	19

Sources Nederlands Instituut voor de Publieke Opinie (NIPO), report nos. 444–447 (13 September 1950); report no. 624 (26 June 1953); and report no. 915 (5 November 1962); Van der Boom (2000: 26–28)[2]

be one without armies, fought by remotely controlled weapons, including atom bombs, assuming the Americans would not be able to keep their nuclear monopoly. Once started, the conflict would automatically escalate, destroying all American and European cities within 24 hours.[3] Dutch newspapers paid a lot of attention to the book *No place to hide* by the American army doctor David Bradley, which was published in the autumn of 1948. Bradley had attended the tests at Bikini Atoll in 1946 and concluded that protection against ionising radiation released by atomic bombs was practically impossible.[4] Widespread fear of nuclear war, in addition to pressure from the USA as mentioned above, also helps to explain the creation of the BB. One MP spoke of an 'SOS of the Dutch people'. The budget, 8 million guilders, to be spent over the next four years, was widely criticised as being much too limited—an 'ostrich policy' as another MP described it (Van der Boom 2000: 36, 115).

During the 1950s, as the nuclear confrontation became an established fact of life, fears were offset by increasing standards of living, and they seem to have abated somewhat. But recurring crises (over the Suez Canal and Berlin, for example) reminded people that nuclear war might break out any time. In 1955, NATO held a large exercise in West Germany, called *Carte Blanche*, which simulated a nuclear surprise attack on Western Europe and tested the response of NATO forces. It involved mock attacks with 335 atom bombs from both sides, mainly on West Germany, but also on a few Dutch targets. Dutch newspapers reported the horrors of this imagined war: West Germany would be devastated, and at least half of its population would be instantly killed. It would be a short and horrible war, a 'mutual suicide', as even the chief of staff of the NATO air forces, Peter Wykeham-Barnes, said. He added: 'Talk of winning such a war is madness', and there would be '[n]o victors, only a few survivors'.[5] It is therefore not surprising that in 1958 the government spoke of a 'latent mood of panic' in the Dutch population (Van der Boom 2000: 150, 221). Another survey, in 1959, showed that people associated nuclear war with 'everything broken, dead, or lost' (Leenders 2001: 150).[6]

The basic mission of the BB was to convince the Dutch population that such pessimism was unwarranted, and that it could, and should, protect itself against aerial attack. This was presented as a civilian task. The name *Bescherming Bevolking* (BB) or 'protection of the population' avoided the word 'defence' because of its military connotation, which planners and PR people knew the Dutch population disliked. For the same reason, the BB was the responsibility of the Minister of Domestic Affairs, not the Minister of Defence. The basic elements of the organisation were local units of volunteers, a scheme called 'organised self-protection'. Citizens were to create shelters in their own homes (see Fig. 6.1).

Those who lived in high-rise apartment buildings should construct a shelter together, preferably in the basement, with some subsidy from the government. In peacetime, these shelters could be used for storage or as parking for bikes. For those caught in a public space at the time of an attack, public shelters would be built at busy junctures in cities. None of these shelters would be able to withstand the heat and shock of a close nuclear explosion, but several miles away from ground zero they would be able to save lives. This basic structure remained in place until the dissolution of the BB in 1986.

Fig. 6.1 Civil defence: The patriarchal view of the BB
Notice how the husband, as helpless as his wife and children, seems to protect them with his left arm, which extends to the stairways. No hint here of the mother as the protector, the idea behind the women'swomen voluntary efforts. The image is strikingly similar to the advertisements featuring families analysed in Goffman (1976: 37–39) and the description of the American situation in McEnaney (2010) (From the booklet *Toelichting op de wenken voor de bescherming van uw gezin en uzelf* [s.l., September 1961], p. 5. *Source* National Archives, The Hague. Archive of Organisatie Bescherming Bevolking, bloc 1960–1969, box 73)

No serious protection against air raids had existed in the Netherlands up to this point. During the 1930s, the government had made a half-hearted attempt to create a mostly voluntary organisation that would fight fires and dig trenches where the population could take shelter during an aerial attack, but people had never taken this seriously, and the government's advice to citizens for individual self-protection was largely ignored—a response that would later be repeated with the BB (Van der Boom 2000: 31–32; Bosma 2012: 76–91).

The BB estimated it would need 200,000 volunteers.[7] The government started a big propaganda campaign in the spring of 1952, with short films, radio talks and pamphlets. It was aimed mainly at men, although women were eligible for the Social Care Service, that was to provide victims with food and comfort, 'work for which a woman's hand was created' (Van der Boom 2000: 51). BB recruiters reported that it was often women who persuaded their husbands to join the organisation. It was also much easier to find women to join the Social Care Service than it was to find men for the other units (Hueting 1995: 104, 107–108; Van der Boom 2000: 51, 63, 67, 69, 134). The overall result of the campaign was disappointing: only 123,000 people signed up. Recruitment therefore continued after 1953, with similar, if not worse results: volunteers had already begun leaving the organisation. This led to the attempt to recruit more women, which will be discussed below. Another way to solve the volunteer problem was to enlist young conscripts. This met with fierce opposition, but at least from the mid-1960s, the ranks were almost filled.

In 1955, the Minister of Domestic Affairs announced what experts had known for years and the population had feared for almost a decade: in the case of a war, the Netherlands would suffer several nuclear attacks, on Rotterdam, Amsterdam, The Hague (seat of the government) and several other strategic spots. The BB was reorganised to deal with this not-so-new threat. Regularly, the government announced that more shelters would be built, but actual building was often postponed or silently called off.[8] In 1968, the Minister of Domestic Affairs reported to Parliament that at that time shelters with fallout protection were available, or under construction, for 66,200 persons; the total population at that time amounted to 12 million.[9] From the late 1950s, an increasing number of experts and politicians expressed their scepticism about the organisation (Van der Boom 2000: 261–270). Prominent Dutch scientists argued that the government grossly underrated the effects of a nuclear attack. The BB's pretence of protection only served to prepare the Dutch population for a nuclear war, they said. The only sensible way to prevent such a war was to completely abolish nuclear weapons.[10]

In the summer of 1961, tensions around Berlin increased, and a nuclear war scare gripped the Western world. The Dutch government, like governments in other West European countries, responded by having a pamphlet delivered to each home, entitled 'Suggestions for protecting your family and yourself'.[11] It explained what to do in the event of a nuclear attack. The main advice was to seek shelter at home or the nearest place available. Suggestions were given for creating a home shelter, what to keep in stock there and how to provide first aid. Three weeks later, an illustrated, 30-page brochure with further explanations was distributed. It emphasised protection against fallout, perhaps because that was more attainable than protection against the shock, heat and immediate radiation of a nuclear explosion (Weart 1988: 258). The overarching theme of this campaign and the booklet was to counter the widespread 'defeatism', the deep pessimism, that had taken root during the 1950s, and to convince the population that self-protection made sense: fatalism, the BB argued, allowed panic to take over and opportunities for survival to be wasted (Van der Boom 2000: 208; on 'panic management', see also the chapter by Bjørnsson in this volume).

The pamphlet and the brochure were not well received. According to a survey held shortly after the campaign, a little over half of the respondents said they had read them, but of those who had, only half could say what fallout was, and an even larger proportion had no idea what to do to protect themselves. A survey held in 1968 showed that hardly anyone had created a shelter at home—not even BB people and politicians (Van der Boom 2000: 232; Valkenburgh 1964: 231)—and municipalities had almost completely neglected their duties in this respect. One example is the small town of Barendregt, close to Rotterdam. The civil servant answering the questionnaire reported that the new municipal office had a shelter facility for the local government. He did not know how many people had a shelter at home, but he supposed there were very few. He announced that Barendregt was preparing further measures. But in 1972, the town reported that for about 12,000 of the 13,000 inhabitants there was no shelter. In short, Barendregt's citizens were not interested in the topic at all, and neither was their municipal government.[12] This was typical of the Netherlands. The mayor of Amsterdam at first even declined to carry out the survey, because he found it meaningless (Van der Boom 2000: 248; Leenders 2001: 227–235).

Some people were angry. They said the BB suggested that a nuclear war could be fought and survived.[13] A well-known young writer, Harry Mulisch, wrote a savage satire, which was published a few weeks after the BB's brochure: 'Suggestions for protecting your family and yourself on Judgment Day'. It alternated horror scenes from the biblical Book of Revelations with suggestions from the BB, including sandbags and ice cubes (Mulisch 1961). Most other reactions published in newspapers were scathing or angry. After this PR disaster, the BB kept a low profile, and in the 1970s, many people wondered if it still existed at all. Opinion polls showed that the number of people expecting nuclear war decreased rapidly during the second half of the 1960s.[14]

Why was the organisation so unpopular? First, neither the central government nor municipal governments believed in civil defence. The budget was kept low and the organisation was set up half-heartedly. During the first few years, BB volunteers had to wear unimpressive overalls as a uniform, local leadership was often weak, courses were uninspiring, and exercises were not taken seriously. Leading figures were often retired army officers who regarded their job as a kind of sinecure. Mayors and councillors commonly showed a complete lack of interest. The BB increasingly became a laughingstock, and its volunteers were ashamed to be seen by their neighbours wearing their uniforms or doing their exercises (see Fig. 6.2).[15]

One result of this lack of funding and leadership was continuous underperformance. The delays in shelter building illustrate this. Most people were unaware of the location of shelters (Van der Boom 2000: 158–159). In practical tasks, the BB also cut a bad figure. When the organisation was called on to assist during emergencies—at a railroad disaster in 1962, at an accident at a chemical plant in 1972 and on a few other occasions—its performance was considered worthless (Leenders 2001: 230–231).

Dutch Women in the Cold War: Social Position, Ideals and Voluntary Work

While the BB became increasingly unpopular, women volunteers set up their own, vigorous campaign to instruct women in self-protection. What motivated them? And how did they deal with the BB? In order to understand this, we have to consider the postwar debate about the role of women in society.

The Second World War shook Dutch society to its roots, including the relations between men and women (De Liagre Böhl 1987; Van Liempt 2009: 265–314, 366–416; Schwegman and Withuis 1993: 568–570). The illegitimacy of the occupying German authorities during the war encouraged illegal activity. This included hiding, and feeding, some 25,000 Jews, and 300,000 non-Jews, a very risky business in which women played a key role. Towards the end of the war, when food and fuel became scarce, pilfering and prostitution spread. After liberation,

Fig. 6.2 BB ridiculed
Self-mockery in the BB magazine *De Paladijn* 1964, no. 10. Notice how marginal women are

300,000 men who had been deported to Germany to work in the factories returned home. Many others had escaped deportation by going into hiding. Women therefore had taken up the task of feeding and holding together their families. When liberation came, many young women had brief affairs with Canadian soldiers, which led to a spike in illegitimate births, divorces and cases of venereal disease a year later. The public humiliation of young women who had been too friendly with German men (*moffenmeiden*) can easily be interpreted as revenge for this afront to Dutch male pride (Judt 2005: 42–43). In other words, the war had not only made women heroines but also put them morally at risk. It had inspired a collective self-confidence in women while diminishing the standing of men.

Small wonder that genuine moral panic broke out among politicians and church leaders after the euphoria of liberation had waned. They believed the country had to restore not only its buildings, harbours and bridges, but even more so its decency and orderly habits. As a journal for mental health put it: 'The sense of mine and thine, and respect for the authority of the government have been severely weakened, pent up aggression has not yet found constructive outlets, and the absence of fathers and the sexual misbehaviour of mothers have created unacceptable situations' (Mol and Van Lieshout 1987: 80). The response to this state of moral emergency was the creation of a range of institutions for social work and therapy which focused on re-establishing traditional family relations. The motto was: 'Restoring the family will restore the nation' (*Gezinsherstel brengt Volksherstel*) (De Liagre Böhl 1987: 28). This was not difficult: it entailed the restoration of the organisational networks created before the war that were dominated by the protestant and Roman Catholic churches which embraced a patriarchal view of society.

This restoration clearly succeeded.[16] Throughout the 1950s and 1960s, most young women married and had children. Divorce was unusual until the late 1960s. Very few women, a much lower percentage than elsewhere in Europe, worked outside the home, in spite of the shortage of labour. Even in 1965, 83% of the respondents in a survey rejected the idea of mothers of schoolchildren working away from home. But there were changes too. The legal position of women began to improve from the late 1940s. For example, both parents acquired equal authority over their children (1947), married women were given the same rights as their husbands to their common possessions (1957), women no longer automatically lost their job upon marriage, and equal pay for equal work was introduced (1957). Many of these rules were routinely evaded, but clearly, the position of women was no longer fixed—it was debated, changes were proposed and sometimes adopted, and even if not fully implemented, they were 'in the air' (Brinkgreve and Korzec 1978: 97–99; Schwegman and Withuis 1993: 575). And many women believed the war had proven that their role in society was essential.

Profound and lasting changes in attitudes started only in the late 1960s, brought about by economic growth, the introduction of the contraceptive pill, and the second feminist wave (Brinkgreve and Korzec 1978: 115, 126–127; Pott-Buter and Tijdens 1998: 139–141; Schwegman and Withuis 1993: 579–581). Women became more highly educated and more eager to pursue a career. Motherhood was no longer idealised; women were now expected to 'develop themselves' beyond the family. The number of children per family started to decline and divorces increased (Van Setten 1987: 25–26; Brinkgreve and Korzec 1978: 114).

The high tide of women's participation in civil defence therefore occurred during the period before the second feminist wave, when a restored patriarchal order collided with a new sense of responsibility and competence among many women. Although feminism is often thought to have almost disappeared during this period, the role of women in society was actually a big topic of debate. Simone de Beauvoir's *Le deuxième sexe* (1949) and Margaret Mead's *Male and female* (1950) were only the most famous of several books on this topic that appeared during these years.[17] In this section, I will sketch the ideas of four widely read Dutch authors and one lesser known one. While different in many ways, they shared key assumptions and concepts about women in society, ideas that also informed the activities of the volunteer organisations and which we can compare with the imaginary of the BB.

My first witness is Martina Tjeenk Willink (1905–1992), a lawyer and the only female member (for the Social Democratic Party) of the senate. At a meeting of the Union of Women's Volunteers in 1948, she gave a speech on 'Woman in leading positions'.[18] The recent war and the current nuclear threat, she says, show that our culture is in deep trouble. This is caused by three potentially valuable tendencies which have run amok: individualism has isolated humans from their deep connections with the world, rationalism has blocked from view the spiritual values hidden in the mysteries of life, and technology, developed to serve us, has enslaved us (she mentions the atom bomb and the assembly line to illustrate this).

All three, according to her, are products of 'centuries of male dominance'. To prevent disaster, we should foster qualities that are 'generally human', but that are more strongly developed in women: connectedness with life, which seeks to bind rather than divide, and which is engendered by our motherhood—whether or not we become mothers ourselves—and a balance of thought and feeling: 'the mind should respect the wisdom of the heart, as the heart should learn the discipline of the mind'. In her reasoning, that is why women should serve 'the human family', the world in which they want to see their children grow up. In her assessment, the recent war has widened women's sense of social responsibility and given them confidence in their capabilities. These are gains that have to be developed further, both for our personal development and for society.

A few years later, the physician-philosopher Frederik Buytendijk (1887–1974) published his influential study *De vrouw* (*Woman*, 1951), framed as a reply to De Beauvoir's book, which he called the *most important* work on the subject (his emphasis) (Buytendijk 1969: 25). *De vrouw* was reprinted 16 times until the late 1960s. Like De Beauvoir, Buytendijk argued within an existentialist framework, but he differed from De Beauvoir and Sartre in emphasising the meaningful relationship of human beings to their world.[19] His core argument was that being human meant being physically in the world. From birth, a person's body was the 'preliminary sketch of his existence', and being male or female invited a certain relationship with the world, without being deterministic. A woman's body invited her to develop her caring qualities, which culminated in motherhood. However, a woman could cultivate all human capacities, including those that were usually associated with men, as had become evident during the war and on many other occasions (Buytendijk 1969: 9). And the gift of caring for life was human, rather than exclusively female. Buytendijk's learned treatise did not differ basically from Tjeenk Willink's short speech: he too spoke of the need for qualities that women can develop more naturally, and he also liked to speak of the 'mysteries' of human existence that could not be grasped by scientific analysis.

Two years before Buytendijk's book came out, in 1949, a stage play, *De wereld heeft geen wachtkamer* (The world has no waiting room), written by Maurits Dekker, caused a small sensation in the Netherlands (Van Lente 2012: 155–156). It portrayed a group of researchers working on a military-funded project to develop a more powerful explosive than plutonium. When a young researcher becomes fatally ill after exposure to radiation, the elderly housemaid of the professor who leads the project observes how the militarisation of research has changed the atmosphere in the team. No longer do the students come and visit the professor, making jokes, she says, but 'bigwigs in uniform who have secrets and suddenly shut up when you enter the room'. The professor, she has noticed, has become tense and sleepless. 'What for?' she exclaims, 'Why don't people just live?' She expressed what Tjeenk Willink and Buytendijk had formulated as a typically female rejection of the destructive, male pursuit of nuclear weapons research. No doubt, the author was on her side and wanted the audience to feel the same.[20]

Perhaps the most profound (and beautifully written) exploration of the role of women in the atomic age was the well-known novelist Hella Haasse's (1918–2011) long essay *Een kom water en een test vuur* (A vessel of water and bowl of fire, 1959). The 'vessel of water and bowl of fire' of the title were the gifts that, in Roman times, the bridegroom offered to his bride as she entered the home of which she was to be the *domina* dressed in an orange and saffron robe, as if she were a flame. The fire stood for erotic passion (the origin of all culture, said Haasse), but also for the oven that magically transformed dough into wholesome food, and of course for her womb, which would produce new life. And water was the symbol of life itself. According to Haasse, the modern age, in which water comes from a tap and light from an electric lightbulb, has lost contact with these elementary phenomena. The awareness of women's deep and mysterious connection with the basics of life is gone. Brides now wear a white dress, symbolising their untouched delivery to the closed-off space of the home where the man rules. This state of affairs is deeply troubling now that water and fire have entered into a monstrous contract in the hydrogen bomb. As humanity is going through its most drastic transformation ever, women should reconnect to this 'cosmic space', symbolised by water and fire. For '[t]he whole world is the home of woman' (Haasse 1959: 17, 77). Their special mission was to help others to reconnect to the larger whole, both in a mystical and in a practical sense: to devote 'all our attention, love, and energy to what is given to us, 'earthy reality', things, human beings, nature' (Haasse 1959: 34–36, 150). This would serve as a counterforce to the male tendency to dominate nature, which had led to the frightening, soulless world of modern machinery.

My final witness, much less well known, is Jo Le Rütte (1888–1972), a retired secondary school teacher, who in 1962 published a pamphlet called *De vrouw en het Atoomtijdperk* (Woman and the atomic age). She was a pacifist, Esperantist and an early member of the Dutch pacifist party, which was founded in 1957.[21] Like Haasse, Le Rütte characterised her time in terms of technology: on the one hand, spectacular material improvement after the war, and on the other hand, the coming of the atomic age, bringing constant fear of utter destruction (Le Rütte 1962: 6). She identified technology with male power: certainly not bad in itself, but potentially deadly. 'If we walk the same road as we do now, humanity will be destroyed by technological violence' (Le Rütte 1962: 29, 31). This can be prevented only by placing international affairs in the hands of men and women together, just as children are brought up by a father and a mother. Like the other authors, she regarded the family as the core of society, and motherhood the core quality of women, endowing them with a caring disposition. This care should be extended to the world at large. She proposed, among other things, giving women prominent positions in economic and financial policy—which were similar to managing the family budget—general disarmament and the spread of Esperanto, so that people around the world could get to know each other (Le Rütte 1962: 42).

Le Rütte's pamphlet reminds us that the maternal drive could lead to pacifist activity as well as to supporting civil defence. Lawrence Wittner has documented the prominent role of women in antinuclear movements in several countries from the mid-1950s onward (Wittner 2000). I have not found much evidence for this in the Netherlands, although members of the pacifist party to which Le Rütte belonged sometimes explained their convictions in these terms. One of them wrote later, during the 'second feminist wave': 'Even if you reject the mother role, you don't have to discard your caring and emotional capacities. After all, we have to teach men to also become more caring and "maternal"' (Denekamp et al. 1987: 68).

These women shared the notion that they had a special, caring task in society. This idea had originated in the first feminist wave, but the war and the threat of nuclear war had increased its urgency: women's participation in public affairs was necessary to stymie destructive male tendencies.

This idea found practical expression in the rise of voluntary movements, especially in the 1930s. In 1938, when international relations were tense and war loomed, a group of ladies (most were from wealthy families and many held academic degrees) established the *Korps Vrouwelijke Vrijwilligers* (Women's Volunteers Corps), which trained its members to help protect the population against the consequences of air raids. The Korps presented itself as 'an instrument for the government when it needs the population to provide help'. This might be in the event of war, a flood or some other disaster (Hueting 1995: 10, 16). Hundreds of women joined the Korps, and very soon there were 30 local chapters all over the country (Hueting 1995: 28).

In May 1940, three days before the German invasion, the Korps offered its services to the National Inspection for Protection against Air Attacks. The government officials rejected the offer. The women, baffled, decided to continue work on their own (Hueting 1995: 37). When the inner city of Rotterdam was destroyed by German bombs a few days later, the Korps provided shelter and food to hundreds of refugees. At the end of the war, girls and women from the Southern part of the country, which had been liberated first, went to help exhausted housewives in the North to clean up and repair their homes. And in 1953, when the South-Western part of the country was inundated during a storm, women volunteers again provided essential help.

In the early postwar years, a conflict arose between the *Unie van Vrouwelijke Vrijwilligers* (Union of Women Volunteers, or UVV), the successor organisation of the Korps, and several local women's voluntary organisations that did not want to be part this organisation. A new umbrella organisation was created in 1951, the *Nederlandse Federatie voor Vrouwelijke Vrijwillige Hulpverlening* (FVVH) (Hueting 1995: 71–86). The UVV declined to join the new federation. The English Women's Voluntary Society, the UVV's admired example, declared that it would only deal with the UVV. This complicated conflict, although not important to our topic in itself, does explain the strange way in which women's voluntary organisations became involved in civil defence a few years later.

Women volunteers and the BB

When the government began recruiting volunteers for the BB in 1953, the UVV and other women's organisations encouraged their members, except mothers of small children, to join. As we saw, many women volunteered for the Social Care Service and prompted their husbands to join the organisation too. Later that year, the Department of the Interior sent three women, including the president of the UVV, to England to learn about civil defence from the Women's Voluntary Service. As the BB struggled to get started, the UVV decided to follow the example of the WVS, which had initiated a campaign in 1955 to educate small groups of women in 'self-protection'. The campaign was called 'One-in-Five', indicating that it aimed to reach one fifth of all adult British women, a goal that seemed more realistic than attempting to reach all women (Hunt 2012). Women were invited in groups of ten to twelve persons, who would meet three times to be instructed on how to minimise the impact of fire, radiation and blast caused by a nuclear explosion, and how to care for the injured. Instruction was provided by a volunteer who had followed a short course.

When the BB began its campaign to recruit more women in 1956, it used the rhetoric of women's natural caring talents.[22] A recruitment folder said that although the BB was not usually associated with women and girls, their female qualities made them perfectly suited to this kind of work. 'All women share the quality of self-sacrificing love', the pamphlet said, and every woman's heart was inclined 'to help, to serve' people in need, as had been demonstrated so well during the flood of 1953. While some tasks were obviously suitable for women (first aid, providing food and shelter, administration, communication), no position in the BB was unattainable for them—including higher staff functions.

As it turned out that new women volunteers were recruited mainly in areas where the BB had hardly campaigned (Van der Boom 2000: 71–72), the organisation asked the FVVH, in February 1958, for help educating Dutch women in '*Individuele Zelfbescherming*' (individual self-protection).[23] The BB official responsible for recruitment explained that while there were now 166,000 volunteers, including 30,000 women, 250,000 volunteers were needed. He calculated that about two million Dutch women would be eligible for the BB—not counting elderly women and young mothers. Because men would have to do military service or be at their jobs in the event of war, women would have to take responsibility for protecting their homes. And 'obviously', he said, the FVVH was needed to recruit these women. He then repeated the formula of women's fitness for all positions in the BB, their natural caring qualities and their sense of responsibility, which had been proven during the Second World War and the flood of 1953. Their participation would also help give the BB a stronger position in society. In this way, they could contribute to defending the West, which was one of the BB's purposes.[24]

The FVVH agreed to help the BB, but then discovered that two women from the UVV were already taking a course on the One-in-Five campaign conducted by the British WVS, and that these ladies were determined to set up a similar programme in the Netherlands. As the UVV ladies' trip was financed by the Department of the Interior, which was also responsible for the BB, this was a curious mistake for the BB to make—and not the last one. The minister, by the way, had approved both the UVV's plan and that of the FVVH.

The UVV's One-in-Five programme was at first directed towards the four large cities in the West: Amsterdam, Rotterdam, The Hague and Utrecht. The mayors of those cities were to determine when and where the instruction meetings would be held. The three-part course was devoted entirely to protection in the case of nuclear war, and all women in these cities would be invited. The FVVH set up a rather different programme. It consisted of a much more loosely structured course about all kinds of accidents that could happen in and around the house ('Safety in the home'). Protection against radiation was only one part of the programme. The course was meant for members of the organisations that belonged to FVVH.

When the board of the FVVH discussed its plans with the two UVV ladies, the latter criticised the looseness of the FVVH course and said that they had already been invited by mayors of other cities, besides those of the four large ones. They were very displeased that the FVVH was setting up a programme of its own. But when the FVVH consulted its member organisations, several of them showed strong resistance against the UVV's approach. They rejected being at the beck and call of the mayors, or of the BB. One woman argued that having completed a course to become an instructor, no mayor could keep her from teaching the members of her organisation. Another said that the members of her organisation did not want to deal with emergencies relating to nuclear war only.[25] And the secretary of the Liberal Protestant Women's Federation wrote to the BB that several women in her organisation were anti-militarist, and therefore rejected cooperation with the BB.[26]

Clearly, One-in-Five was much more in line with the BB's purposes than the FVVH's plan. Like the BB (and like its British example, the WVS), the UVV tried to convince women that a nuclear war was not likely to destroy the whole country, and that preparing for an attack strengthened the defensive stance of the country. It also argued that the population's attitude towards the BB should be bent in a more positive direction.[27] The Minister of the Interior supported the UVV. He wrote to the FVVH that instruction meetings should be organised only upon the request of the mayor, and that the FVVH did not hold a monopoly on this activity.[28]

Both the UVV and the FVVH seem to have realised that in order to be successful, they should keep their distance from the BB, as that organisation was unpopular with the wider population. For example, when the BB proposed providing the FVVH with pre-printed meeting invitations which also contained BB slogans, the FVVH replied that they preferred other means of reaching their members. Aversion to the BB was expressed frequently in meetings. It must have been aggravated when BB made another tactical mistake: it distributed its brochure 'Suggestions for protecting your family and yourself' (September 1961) without consulting, or even informing, the women's organisations.

In 1960, the two organisations managed to reach a compromise and created one programme, which was rather similar to the One-in-Five plan. The FVVH would coordinate the project. Two years later, the FVVH had managed to recruit and train 1,600 volunteer instructors.[29] The campaign proceeded smoothly and was rather successful. The wives of mayors played an important role in propagating the meetings. Instructors often reported the lively interest of their audiences and requests from several organisations to organise courses for their employees.[30] The 'Suggestions' brochure led to a short boom in requests for meetings, but the ridicule of the BB that followed must have harmed the demand for courses too. There was another brief increase of interest in 1965, when the FVVH organised meetings following a course on television about accidents at home (not specifically nuclear war). 4,577 women participated. But in January 1966, fatigue set in. One of the main organisers said that most people did not want to hear about war, and that she had become averse to giving courses. Many women said they felt the same way.[31] By that time, 43 municipalities had attained the target of reaching one-in-five women. Most of these were in the countryside. In the cities, interest was much lower (for variations in the urban–rural divide in Britain and Sweden, see the chapter by Cronqvist and Grant in this volume). Overall, only 12% of the number aimed for was reached. Still, 88,594 women had taken the FVVH's course.[32]

In 1965, the government learned from a preliminary survey that, for many people, creating a shelter in their homes was more difficult than had been assumed. It was decided to investigate precisely how many shelters actually existed (Van der Boom 2000: 157–167). The FVVH was asked in 1966 to suspend its courses until the results were known. This took several years. The Minister of the Interior asked the FVVH to maintain its corps of volunteers in the meantime, so that they could pick up their activities as soon as the government knew more about the shelter situation in the Netherlands. The FVVH responded by organising regular meetings for its instructors, with invited speakers who discussed the latest developments in international politics.[33] When the survey was finally completed in 1971, the minister informed the FVVH that its courses were no longer needed, but that he would appreciate the organisation staying on standby.[34]

Conclusions

It is evident that the FVVH worked much more effectively towards preparing the population for nuclear attack than the BB. Hundreds of women volunteered to become instructors; they often reported successful and well-attended meetings, and were invited to give presentations by several organisations. Although they did not meet the 'one-in-five' target, they did reach a large number of Dutch women. That the FVVH and the UVV managed to overcome fundamental differences in approach and worked out an effective programme together was another indication of their creativity and tenacity. Even when the BB decided to discontinue the courses in 1966, the FVVH kept its corps of instructors up to date on international affairs and the risks of nuclear war, so that they would be well prepared in case the government requested their assistance again.

All of this scores very favourably when compared to the failed recruitment efforts of the BB, the sluggish participation of members in their exercises and the public relations disaster of the 'Suggestions' folder. The FVVH could probably have been even more successful if the BB had not operated so clumsily, first overlooking the government-supported initiative of another organisation, and then publishing and distributing a brochure without consulting the women's organisations first. Its bad reputation must have made the work of the women more difficult, which is why they kept their distance from the BB.

The relative success of the FVVH can be attributed partly to its solid philosophy, as expounded by Tjeenk Willink, Haasse and Le Rütte, that fostered a conviction that the BB lacked. It was an elaboration of the 'maternalist ideology' of the first feminist wave, made more urgent by the threat of nuclear war. It received broad recognition, for example through professor Buytendijk's widely read book, and it inspired women to provide crucial services at the beginning and the end of the war and during the 1953 flood. Another factor was that the FVVH was a grass-roots organisation. A sociological survey carried out in 1962 showed that most Dutch people felt they had no influence on political decision making, and that in international affairs, the Netherlands was simply powerless. The debacle with the 'Suggestions' brochure in 1961 must have reinforced this scepticism towards the state (Valkenburgh 1964: 208, 229–231, 239).[35] Given such a lack of confidence in state authorities, an organisation such as the FVVH, run by women throughout the country, which had proven its competence and organisational power, could much more easily recruit volunteers and keep them engaged. The same survey showed that women tended to be more nationalistic than men and more supportive of military defence, which can be interpreted as indicating their willingness to serve their compatriots in times of a national emergency (Valkenburgh 1964: 239).[36] And finally, as was the case in West Germany (see the chapter by Molitor in this volume), including preparations for a nuclear war in a range of first-aid activities around the house made it much less controversial than, say, shelter building.

Several features of Jasanoff's description of sociotechnical imaginaries clarify this analysis. First, contestation was a conspicuous aspect of the development of civil defence imaginaries (Jasanoff 2015b: 323). All conceptions of civil defence had to be articulated in the face of fierce criticism and scorn, which was basically similar, though not equally intense, in all countries examined in this book. We can add that very different practices could be legitimated with the same maternalist imaginary: not only participation in civil defence, but antinuclear activism and *rejection* of civil defence as well. Second, the women's organisations *performed and enacted* this imaginary, rather than proclaiming and expounding it in articles, books or other documents. Theirs was 'a continually rearticulated awareness of order in social life and a resulting commitment to that order's coherence and continuity' (Jasanoff 2015a: 26). To be sure, this awareness was conceptually well-grounded, and expounded in the work of well-known authors such as Haasse and Buytendijk. It had also proven its practical efficacy. But in spite of all this, the women volunteers failed to undermine the dominant paternalist imaginary embodied in the BB. Jasanoff writes that a new imaginary can only assert itself if 'an imagined new order draws on deeper notions of how societies ought to fit together', or when they 'fall in line with the way things are remembered as being' (Jasanoff 2015b: 339). Neither was the case in the Netherlands until in the late 1960s. Only then did the so-called second feminist wave launch its attack against the paternalism that was so deeply ingrained in society. This illustrates how entrenched 'infrastructures' of thought and feeling—in this case gender conceptions and (dis)trust of government—profoundly co-determine sociotechnical imaginaries.

Acknowledgements I would like to express my profound gratefulness to the people who helped me in this project: A. Le Rütte-Veringa and Iris Le Rütte, niece and grand-niece of Jo Le Rütte, who shared their reminiscences of their aunt Jo with me; Roselle Servage and her colleagues, who received and helped me at the Atria archive, even while the rooms were being renovated; Manuel Kaal at Kantar Group in Amsterdam, for locating and sharing NIPO polls with me; Jop Ewijk and Rolf Boost for helping me with the copyrights of Wim Boost's cartoon; and the editors of this volume for their excellent direction of this collective enterprise.

Notes

1. My description of the Dutch civil defence organisation relies heavily on Van der Boom (2000).
2. After 1962, the NIPO stopped regularly asking this question, which indicates a degree of relaxation. But fears could flare up. After the Soviet invasion of Czechoslovakia in August 1968, about half the population believed that war had become more likely, while about 30% said this likelihood had not changed (NIPO report, no. 1229, 12 September 1968). The exact date of the polls matters a lot, as is clear from the data on 1950 (shortly after the start of the Korean war) and 1962: the first poll was taken during tensions around Berlin, but before the Cuban missile crisis started, the second when the crisis was resolved. The date for 1953 is difficult to read. Additional data and analysis in Van Lente (2012). My findings contradict those of Van der Boom (2000: 27, 56, 105) and others who state that the Dutch population only really began to worry about nuclear war from about 1955.
3. *Panorama* 8, November 1946. Based on *Life magazine*, 20 August 1945 and 19 November 1945.
4. E.g. *Volkskrant*, 5 October 1948, *Overijssels Dagblad*, 9 October 1948, *Gooi-en Eemlander*, 13 December 1948, *Volkskrant*, 24 September 1949. Bradley criticised: *Volkskrant* 14 May 1949.

5. A sample: *De Tijd*, 19 July 1955, *Volkskrant*, 9 July 1955, *Leeuwarder Courant*, 30 June 1955, *Trouw*, 29 June 1955, *Vrije Volk*, 29 June 1955. Wykeham-Barnes quoted in *Het Vrije Volk*, 29 June 1955. 'Atom bombs' on the Netherlands: *Trouw*, 25 June 1955.
6. That this was the opinion of most Dutch people was often stated in newspapers, without empirical support: it was an 'everyone knows' type of statement, e.g. *Algemeen Handelsblad*, 31 July 1965.
7. Internationaal Archief voor de Vrouwenbeweging (IAV) in Atria (hereafter: Atria), Archief FVVH, inv. no. 129, L. Tielenius Kruythoff, responsible for recruitment of volunteers, in an explanation about 'More women in the B.B.' ('Meer vrouwen in de B.B.'), November 1957.
8. E.g. *Het Vrije Volk*, 16 February 1962.
9. *Nieuwsblad van het Noorden*, 28 May 1968.
10. Van der Boom (2000: 195–196, 200, 209). For criticism of civil defence in West Germany, Switzerland, Sweden and Britain, see the chapters in this volume by Molitor (Chapter 3), Marti (Chapter 8) and Cronqvist and Grant (Chapter 9).
11. *Wenken voor de bescherming van uw gezin en uzelf* (September 1961), *Wenken voor Eerste Hulp* (September 1961); *Toelichting op de wenken voor de bescherming van uw gezin en uzelf* (October 1961), all in my own collection. See the chapters in this volume by Bjørnsson (Chapter 2) and Bennesved and Sylvest (Chapter 5).
12. Rotterdam City Archives, 1438 Gemeente Barendrecht, no. 2973: Vragenlijst civiele verdediging, naar de toestand op 1 januari 1968; and no 2975, 'Fallout-beschermingsplan gemeente Barendrecht' uitgave 1972; *De Telegraaf*, 4 March 1966.
13. Cf *Volkskrant*, 1 October 1961.
14. Nederlands Instituut voor Vredesvraagstukken (1975: 201, 356–357).
15. Van der Boom (2000: 55–56, 75–86). *De Paladijn*, the organisation's monthly magazine, constantly had to defend itself against criticism and scorn, including from its own volunteers. See the description by a volunteer going to an exercise in *De Paladijn*, 1964, no. 5. More in *De Paladijn*, 1964, no. 4).
16. Goudsblom (1967: 46–49, 128–139), Van Setten (1987: 24–31, 50, 59, 65, 71–72), Schuyt and Taverne (2000: 274–278, 283), and Pott-Buter and Tijdens (1998: 27–28, 136).

17. Others were Helene Deutsch, *Psychology of women* (2 vols., 1944, 1945), Viola Klein, *The feminine character* (1946), Georgene Seward, *Sex and the social order* (1946) and Gina Lombroso, *L'âme de la femme* (1947). For the intellectual context, see Schuyt and Taverne (2000: 384–386).
18. Published in *Raad en daad. Maandblad van de Unie van Vrouwelijke Vrijwilligers*, 3/7–8 (1 July 1948).
19. Here Buytendijk followed Gabriel Marcel and Maurice Merleau-Ponty, who also were critical of Sartre's view of the meaning of life.
20. The young researcher dies after he has sacrificed himself by entering the reactor core and, to prevent an explosion, manually inserting the cadmium regulators which had jammed. In the light of the common symbolism discussed in this section, one could read this as the young man entering a 'counter-womb', which brings death instead of life.
21. She writes a little about herself in a second edition of her booklet *Leven in het Atoomtijdperk! Hoe? Drie brieven naar aanleiding van "De vrouw in het atoomtijdperk"* s.l.s.a. [1968]. Her niece A. Le Rütte-Veringa and her grandniece Iris Le Rütte provided additional information.
22. Stadsarchief Rotterdam, no. 537, archive of the city district head ('wijkhoofd') J.M. Coppens, box 2/4, folder 'Wegwijzer BB 1', recruitment folder 'Ook vrouwen in de B.B.' s.a. [1956].
23. Overviews of FVVH's information campaign in *De Paladijn*, September 1962, 1964 no. 1, 1965 no. 9, and 1966 no. 11; and in Atria, Archief Federatie voor Vrouwelijke Vrijwillige Hulpverlening (hereafter Archive FVVH), inv. no. 122, 'Chronologische gang van zaken mbt Door Vrouwen Aan Vrouwen', 1959; and inv. no. 128, 'Verslag van de actie Door Vrouwen Aan Vrouwen – Een op Vijf', s.a. [1966]. The discussion below is also based on documents in this archive, inv. no. 121–129.
24. Archive FVVH, inv. no. 129, Tielenius Kruythof, speech for FVVH, 16 May 1957; and explanatory letter to FVVH member organisations, November 1957; speech 23 November 1957.
25. Archive FVVH, inv. no. 122, 'Chronologische gang van zaken.'
26. Archive FVVH, inv. no. 129, letters of 20 June and 27 September 1958. Similarly in Britain, the movement against nuclear weapons,

CND, urged WVS to stop the One-in-Five campaign, which it saw as war preparation. See Hunt, 'One-in-five'.
27. Archive FVVH, inv. no 123, Note by Mrs Ras 'Essentiële punten van het 'Een op Vijf' plan', s.a. [1960].
28. Archive FVVH, inv. no. 123, Letter from Minister Toxopeus to FVVH, 5 November 1959.
29. *De Paladijn*, September 1962.
30. One example: Archive FVVH, inv. no. 127, report by a woman instructor about meetings in The Hague, April–May–June 1966, about courses, among others, at a girls' secondary school and for a group of nurses.
31. Archive FVVH, inv. no. 124, report on meeting 7 January 1966.
32. Statistics in Archive FVVH, inv. no. 128, 'Verslag van de actie', s.a. [1966]. And in *De Paladijn,* September 1962.
33. Archive FVVH, inv. no. 124, 128: documents on the suspension of Voor Vrouwen aan Vrouwen Een op Vijf, 1966. On the 'contact days' for instructors during the following years, see inv. no. 125.
34. Archive FVVH, inv. no. 126, report on consultation of FVVH with the Minister of the Interior, 28 October 1971.
35. This study was based on a survey among a random sample of 183 men and 173 women over 23 years of age in the northern city of Groningen in 1962.
36. This is similar to Great Britain shortly before the Second World War. See Noakes (2012: 748).

References

Unpublished Sources

Internationaal Archief voor de Vrouwenbeweging, Atria, archief van de Federatie voor Vrouwelijke Vrijwillige Hulpevrlening (FVVH).
Kantar Group Amsterdam, Archive of the Nederlands Instituut voor de Publieke Opinie (NIPO).
National Archives, The Hague. Archive of Organisatie Bescherming Bevolking.
Stadsarchief Rotterdam, no. 1438, Archive of the municipality of Barendrecht.
Stadsarchief Rotterdam, no. 537, Archive of the City District Head (Wijkhoofd) J.M. Coppens.

Newspapers and Periodicals

De Paladijn (1961–1970).
https://www.delpher.nl/: newspaper database.
Panorama (1946–1961).

Literature

Bosma, Koos. 2012. *Shelter city: Protecting citizens against air raids*. Amsterdam: Amsterdam University Press.

Brinkgreve, Christien, and Michel Korzec. 1978. *Margriet weet raad: Gevoel, gedrag, moraal in Nederland, 1938–1978*. Utrecht: Aula.

Buytendijk, Frederik J. J. 1969. *De vrouw: Haar natuur, verschijning en bestaan*. Utrecht: Het Spectrum, 16th reprint (first edition: 1951).

Dekker, Maurits. 1949. *De wereld heeft geen wachtkamer: Toneelspel in drie bedrijven*. Hoorn: s.n.

De Liagre Böhl, Herman. 1987. Zedeloosheidsbestrijding in 1945: Een motor van wederopbouw. In *In fatsoen hersteld: Zedelijkheid en wederopbouw na de oorlog*, eds. Hansje Galesloot and Margreet Schrevel, 15–28. Amsterdam: SUA z.j.

Denekamp, Paul et al. 1987. *Onstuimig maar geduldig: Interviews en biografische schetsen uit de geschiedenis van de PSP*. Amsterdam: De Bovenbouw.

Goffman, Erving. 1976. *Gender advertisements*. London: Macmillan.

Goudsblom, Johan. 1967. *Dutch society*. New York: Random House.

Haasse, Hella S. 1959, *Een kom water, een test vuur*. Amsterdam: Moussault.

Hueting, Ernest. 1995. *Vrijwillig: Een halve eeuw UVV*. Zutphen: Walburg Pers.

Hunt, Jennifer. 2012. 'One-in-Five.' Publication of the Royal Voluntary Service. https://www.royalvoluntaryservice.org.uk/Uploads/Documents/About%20us/one_in_five.pdf [accessed 13 December 2020].

Jasanoff, Sheila. 2015a. Future imperfect: Science, technology and the imaginations of modernity. In *Dreamscapes of modernity: Sociotechnical imaginaries and the fabrication of power*, eds. Sheila Jasanoff and Sang-Huyn Kim, 1–33. Chicago: Chicago University Press.

Jasanoff, Sheila. 2015b. Imagined and invented worlds. In *Dreamscapes of modernity: Sociotechnical imaginaries and the fabrication of power*, eds. Sheila Jasanoff and Sang-Huyn Kim, 321–341. Chicago: Chicago University Press.

Judt, Tony. 2005. *Postwar: A history of Europe since 1945*. New York: Penguin.

Kool-Smit, Joke. 1967. Het onbehagen bij de vrouw. *De Gids* 1967 II, 267–281.

Leenders, Rieke. 2001. *Als de hemel valt: Bescherming Bevolking tussen fantasie en werkelijkheid*. Tilburg: Tilburg University, dissertation.

Le Rütte, J. 1962. *De vrouw en het atoomtijdperk*. Monnickendam: Nimo.

Mol, Annemarie, and Peter van Lieshout. 1987. Geestelijke hygiëne en de wereldvrede. In *In fatsoen hersteld: Zedelijkheid en wederopbouw na de oorlog*, eds. Hansje Galesloot and Margreet Schrevel, 75–88. Amsterdam: SUA.

Mulisch, Harry. 1961. *Wenken voor de jongste dag*. Amsterdam: De Bezige Bij.

Nederlands Instituut voor Vredesvraagstukken. 1975. *Buitenlandse politiek in de Nederlandse publieke opinie: Inventaris van in Nederland in de periode 1 januari 1960 tot 1 januari 1975 gehouden onderzoek naar meningen en attituden t.a.v. de buitenlandse politiek* (s. l.).

Noakes, Lucy. 2012. 'Serve to save': Gender, citizenship and civil defence in Britain, 1937–1941. *Journal of Contemporary History* 47: 734–754.

Pott-Buter, Hetty, and Kea Tijdens. 1998. *Vrouwen: Leven en werk in de twintigste eeuw*. Amsterdam: Amsterdam University Press.

Schuyt, Kees, and Ed Taverne. 2000. *1950: Welvaart in zwart-wit*. Den Haag: Sdu.

Schwegman, Marjan, and Jolande Withuis. 1993. Moederschap: van springplank tot obstakel: Vrouwen, natie en burgerschap in twintigste-eeuws Nederland. In *Geschiedenis van de vrouw, deel 5: de twintigste eeuw*, eds. Georges Duby and Michelle Perrot, 557–583. Amsterdam: Agon.

Valkenburgh, Paul. 1964. *Mensen in de koude oorlog: Sociologische bijdrage tot onze kennis van internationaal-politieke conflictsituaties*. Meppel: Boom.

Van der Boom, Bart. 2000. *Atoomgevaar? Dan zeker B.B. De geschiedenis van de Bescherming Bevolking*. Den Haag: Sdu.

Van Lente, Dick. 2012. Nuclear power, world politics, and a small nation. In *The nuclear age in popular media*, ed. Dick van Lente, 149–173. New York: Palgrave MacMillan.

Van Liempt, Ad. 2009. *De oorlog*. Amsterdam: Balans.

Van Setten, Henk. 1987. *In de schoot van het gezin: Opvoeding in Nederlandse gezinnen in de twintigste eeuw*. Nijmegen: SUN.

Weart, Spencer R. 1988. *Nuclear fear: A history of images*. Cambridge, MA: Harvard University Press.

Wittner, Lawrence S. 2000. Gender roles and nuclear disarmament activism, 1954–1965. *Gender & History* 12: 197–222.

CHAPTER 7

Ruins of Resilience: Imaginaries and Materiality Imagineered and Embedded in Civil Defence Architecture

Rosanna Farbøl

In April 1951, Director General of the Danish Civil Defence Directorate (DCDD) Arthur Dahl went on a study trip to the UK. During the course of his stay, he visited the British Civil Defence technical school at Eastwood Park, near Bristol, where he was particularly thrilled by a unique feature: a village that looked like it had been destroyed by atomic bombs. There were bombed-out houses, streets blocked by debris, a crashed airplane and an overturned truck. In these authentic surroundings, civil defenders were trained in basic rescue, firefighting, debris clearance, handling of gases and first aid.[1] The debris and rubble of this ruined village gave 'mass and solidity' (Jasanoff 2015b: 322) to the imaginary and dystopian war civil defence prepared for. Upon his return to Denmark, Dahl declared that a similar facility, a town of and in ruins, was 'absolutely necessary' in Denmark, at least in the capital, Copenhagen, to

R. Farbøl (✉)
Department of History, University of Southern Denmark, Odense, Denmark
e-mail: rosfa@sdu.dk

M. Cronqvist et al. (eds.), *Cold War Civil Defence in Western Europe*,
https://doi.org/10.1007/978-3-030-84281-9_7

improve the education of civil defence instructors.[2] Realistic ruins were needed to make the instructors capable of performing their vital skills under the direst circumstances, a nuclear war. In this Cold War vision, ruins were key to resilience and national survival.

From the outset of the Cold War, it was generally recognised that in the next war civilians and not just military installations would be targets of atomic bombs. Cities were considered particularly vulnerable. In consequence, the Danish welfare state sought to protect its citizens through a system of civil defence and emergency planning—in return, of course, for the loyal collaboration of the citizens and on the assumption that the citizens themselves contributed to basic civil defence (Bjørnsson et al. 2020). Undoubtedly, the nuclear condition placed this social contract under stress; particularly with the advent of the H bomb, the state was, in fact, unable to offer total protection from total nuclear war. Still, the ruins could demonstrate to civil defenders and ordinary citizens alike how plans for protecting the population in the imagined aftermath of a war were real: they were based on the material realities of built structures of everyday environments.

Drawing on ruinology literature[3] and on scholarship associated with New Materialism, this chapter focuses on the ruin as both physical form and representation in the analysis of the Danish civil defence's preparation, performance and imagination of a Third World War. The chapter contends that the ruin 'imagineered' (Monteyne 2011) a stage for enacting and performing a future war, whereby the merely imagined was given a concrete, tangible expression. The ruin town added materiality, spatiality, realism and presence to what was largely speculative, discursive and imaginary.

This chapter sketches the transnational extension of the British model ruin to Denmark and examines how the ruin town travelled to a new national context and was appropriated to local needs (and restraints). In Denmark, the ruin town embedded a sociotechnical imaginary that emphasised resilience, survival and regeneration, where welfare triumphed over warfare. Then, disrupting the ostensible linearity in the 'life cycles' of sociotechnical imaginaries (see the Introduction to this book), the chapter demonstrates that the ruin town was re-extended all over the country through a standardisation of ruins that fostered homogeneity and uniformity to become part of the common Danish civil defence landscape, while simultaneously serving to re-embed the imaginary in local contexts. The chapter, finally, discusses whether these ruin towns contributed to an affirmation of social norms and values, arguing that they, in fact, caused a 'taming' (Clark 1999) of the nuclear catastrophe, as well as reflecting and reinforcing a specific political and historically situated understanding of social order and the 'good society'.

Ruins and Catastrophe Imagineered

The ruin is an ambiguous structure that can signal both the loss and the endurance of the past. As physical structure and representational form, the ruin complicates intellectual dialectics between absence and presence, fragment and whole, universal and specific, utopia and dystopia. Gazing at ruins, we not only see the actual structure but also draw on a transhistorical iconography of decay and catastrophe, a 'vast visual archive of ruination' (Hell and Schörle 2010: 1). Though ruins are usually seen as tokens of darkness, death and degeneration, they are also associated with romanticism, gothic tales and the sublime, and taken as heralds of hope, new beginnings and the ability of humans to shape the future (Edensor 2005; Eshel 2010). The Cold War ruin town must be understood as embodying both: it represents abyss and redemption, obliteration and resilience. It offered a spectacular stage for the 'performative dimensions of society's self-reproduction' (Jasanoff 2015a: 5), a stage for enacting, or, we might say, *pre-enacting* (Sylvest 2021), the ability of society to overcome the war and extend existing ways of life into a postwar future.

In 1984, Derrida famously stated that nuclear war was 'fabulously textual' (Derrida 1984: 23): it depended on structures of information and communication because a nuclear war had not taken place. It was, and still is, a non-event (the atomic bombs over Japan in August 1945 ended a conventional war, but they did not spark a nuclear one). The ruin town, however, gave the fiction of war a materiality, a presence beyond discourse. Architectural historian David Monteyne has coined the phrase 'imagineering' to describe the merging of concrete, tangible engineering and architecture with fantasy and imagination (Monteyne 2004, 2011). The ruin town is an often overlooked yet striking example of the apocalyptic city imagineered. It represented the ordinary, everyday townscape rendered into a violent, catastrophic urban battlefield. These civil defence environments made it possible to rehearse and enact the imaginary war in realistic, palpable settings and provided immediate physical as well as sensual access to experiences that seemed to defy representation.

The construction of ruins was a way of mitigating disaster. The ruin town signalled a belief that material structures, the built environment and technical skills could offer protection (Deville et al. 2014), while it also in itself constituted a materialisation of the conception of risk. It is important to keep in mind that it is not just the material structure as such that matters but the co-production (Jasanoff 2015a, b) of the ruin and the ideas, fears and ambitions that turn imaginaries into practice and comes to constitute a social reality. The townscapes were, on the one hand, a result of the imagination of civil defence officials and technicians: they literally set in stone the official civil defence sociotechnical imaginary of what a future war would look like and how it could be prepared for based on the capabilities of the organisation. On the other hand, it in turn came to define and structure the imagination, fostering a certain inertia and immutability.

In highlighting the material powers of the ruin towns, I am inspired by approaches of New Materialism that attempt to overcome 'the enduring separation between thing and meaning, concrete and abstract, physical and mental' (Ireland and Lydon 2016: 2) by shifting analytical focus from what things *mean* to what they *do*. These insights are useful to substantiate and operationalise Jasanoff's notion of materiality, which often appears simply to mean technology (Jasanoff 2015a: 8, 20, 22; 2015b: 324). The concept of 'the material' in this chapter refers in an intentionally ambiguous manner to both the material substances (the built structures and infrastructures of the ruin town) and 'the palpable power effects flowing from the strategic arrangement of material objects' (Frohman 2011). Like shelters or scale models, ruin towns not only reflected risk but also produced, performed and transformed risk and preparedness by materialising statements of the plausibility of war and concretising the idea of state-led preparedness and protection (Deville et al. 2014; Puff 2010). The buildings not only framed but also structured the anticipated event. Ironically (but following the logic of co-production), in the attempt to provide technological solutions for an attack feared likely or even inevitable, the architects of the ruin town contributed (psychologically) to the very construction of the probability or inevitability of this event.

The architect Eyal Weizman suggests that 'the ruin has an 'architecture' in which controversial events and political processes are reflected and from which they might be reconstructed and analysed' (Weizman 2011: 111). While Weizman is concerned with the authentic ruins of Gaza, his approach of 'forensic architecture' (Weizman 2012) can be employed in the study of the artificial ruin town because it is not just an illustration of (anticipated) violence but a source of knowledge about historical tensions, social norms, expectations and fears. As an embedding of a sociotechnical imaginary, it is not innocent but must be understood in a specific historical, social and political context that simultaneously shaped and was shaped by rationalities, behaviours and imaginations of future (and past) war(s).

I have combined these theoretical approaches with original field investigations in a selection of ruin towns still in existence (and use) in Denmark. I have analysed and compared the ruins, interiors and exteriors, as well as the spaces between the ruins, the infrastructures and general townscapes. I have consulted the Danish National Archives, local archives and municipal building construction archives to track down blueprints, situation plans, and technical and statistical calculations, as well as correspondence between the DCDD, local civil defence, contractors, architects and engineers. Finally, I have employed a wide array of national and regional newspapers in order to examine the impact of the ruins on national and local communities and identities.

Getting to Denmark

A couple of months after Arthur Dahl had returned home from his visit to the UK, he sent the DCDD's preferred architect F. Grut and Brigadier Chief O. P. Petersen to inspect the technical school at Eastwood Park with the aim of establishing a similar structure in Denmark. They were received by architect Lewis E. Trevers, Home Office Consultant for technical schools. Trevers suggested that a Danish training ground be initially constructed as a set of few buildings with multiple purposes instead of multiple, single-purpose buildings like Eastwood Park. This would reduce the estimated costs to around DKK 300,000–400,000. According to Trevers, the total construction costs of Eastwood Park had been approximately DKK 1,200,000.[4] A gradual expansion could then take place every year with financial allocations on the national budget.

Both Grut and Petersen made reports following their visit, which make it possible to trace their journey and impressions. Clearly amazed by the facility, Petersen described how training was conducted in the village and specified which civil defence functions (including rescue, work in confined spaces, reinforcing buildings) each of the buildings served.[5] Grut's report was more technical in nature. Focused on the professional construction of realistic ruins, he detailed how the more or less undamaged part of a building was constructed in an ordinary fashion. Then, debris was added based on a combination of empirical knowledge and technical calculations. Grut made an apt observation when he remarked that '[t]he planning of the construction as a whole, moreover, demands a not inconsiderable display of imagination in order to obtain as realistic an effect as possible'.[6] Back in Denmark, architect Grut set to work designing a training ground based on his notes, photographs and sketches. However, a lack of funding meant that Dahl and Grut's ambitious plans for a ruin town near Copenhagen were shelved. The idea, nevertheless, was realised elsewhere, where there were fewer bureaucratic restraints on the display of creativity.

At the civil defence barracks in Tinglev, a village in the countryside of Southern Jutland (about as far from Copenhagen as possible), section leader K. M. Nielsen set about acquiring a bombed-out village—or ruin town as it became quickly known—for the training of his detachment of the South Jutland mobile column.[7] With assistance from platoon leader (and architect by training) L. Weisdorf, a blueprint was drawn up (Fig. 7.1).

The ruins were built between 1952 and 1955 by conscript personnel in their spare time using only free materials from houses designated for demolition in the vicinity.[8] Initially, the ruin town contained four blocks of houses consisting of six one- to two-storey houses, including a shop and a church-like tower, as well as some wooden sheds. The houses were located around a main street (*Storegade)*, a parallel street called Fire Avenue (*Brandallé*) and two smaller streets across. As a whole, it represented a fragment of an urban environment. More buildings were soon built, including a three-storey factory, and a short stretch of railway tracks was added as well. As it turned out, the approach was not unlike the one Trevers had originally suggested. The streets were blocked by rubble and debris, as well as crashed cars, buses and trams.[9] Telephone poles, sewers, electricity and gas were installed, so personnel would also have to take this utility infrastructure (or, rather, broken infrastructure) into account during the exercises. The recruits would face burning houses, smoke or gas-filled basements, and there would be live casualties to be rescued from the buildings. A loudspeaker installation was used to make sound effects, such as of air raids.[10] The exercises often took place after sunset, and the scene was set for an inferno: a town ravaged by firestorms and reduced to

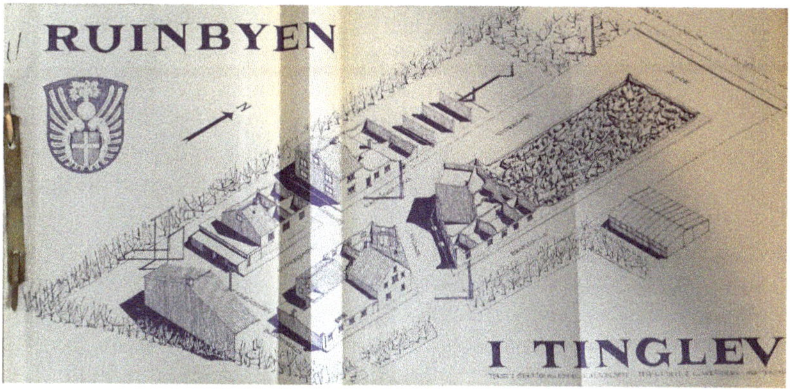

Fig. 7.1 Plan of Tinglev
(Original plan of Tinglev as envisioned by K. M. Nielsen and L. Weisdorf. The factory looks out over Main Street (Storegade), which is outlined by houses in various stages of ruination. To the North-East is an area of concrete debris that were used for practising how to run with stretchers and victims across impassable terrain. Printed with permission from S. Weisdorf)

ruins, the screams of victims filling the night. Training in the ruin town was a corporeal and sensory experience as well as a practical one.

According to Arthur Dahl's successor as director of the DCDD, Erik Schultz, Tinglev ruin town was not perfect from an educational perspective because it was constructed on an ad hoc basis instead of according to Grut's original master plan. Still, it was functional and, significantly, cheap. Schultz estimated the costs at a few thousand DKK, though he stressed that, in effect, it was worth far more.[11] The cost efficiency was also noted smugly in a newspaper article in *Horsens Social-Demokrat*, which compared Tinglev to the British prototype. Exaggerating a good deal, the paper asserted that they were 'similar in size and construction'.[12] Tinglev ruin town quickly became a national source of pride. An article in the national newspaper *Jyllands-Posten* described it as 'an outdoor school room, where it is possible to create all the situations that will arise if the destruction of a war hits the civilian population in an urban area',[13] revealing an expectation that due to the ruin town, the future had become knowable and possible to prepare for: this was what war would look like in Denmark.

Danish newspapers celebrated Tinglev as unique (on the European continent) and revelled in the fact that visitors came from all over the country and abroad on civil defence study trips. The guests were most often from Scandinavia, but also from the other Nordic countries, as well as Germany and the UK. More exotic visitors included Saudi and Thai delegations.[14] The ruin town quickly grew to be the largest training ground in Northern Europe—and, at the time of writing, it still is—and the narrative changed from Tinglev being modelled on Eastwood Park to Tinglev being a template for others.

The success of the ruin town must account for the fact that Tinglev was chosen to become the (only Danish) Civil Defence Technical School in 1956. Instead of being a training ground for the local mobile section, Tinglev was now a national school where conscripts and volunteers from all over Denmark came for technical education and practical training. That meant considerable development and extension of the main building as well as the ruin town. Though initially sidelined, the DCDD now took charge. Architect Grut was called upon to do the work, and it appears likely that some of his original drawings may have been put to use.[15]

Building the Perfect Ruins

During the 1960s, ruin towns mushroomed all over the country. A revision of the Danish Civil Defence Act in 1962 introduced conscript personnel to local civil defence in provincial towns, something local authorities had long called for due to problems attracting sufficient numbers of volunteers. With the sudden influx of large numbers of conscripts, a need arose for better educational facilities locally. In 1962, the DCDD decided to spend initially DKK 1.9 million on building ruin towns. Six larger provincial towns were selected as the first to receive funding for ruin building: Aalborg, Randers, Holstebro, Esbjerg, Århus and Odense.[16] Just a decade later, there were 26 ruin towns spread across the seven Danish civil defence regions.[17] It is not without irony that at the same time as geopolitical détente thawed the Cold War climate internationally, the imaginary war materialised in earnest as an urban reality in Denmark.[18]

Local authorities perceived ruin towns as vital to civil defence training and were eager to start building their own. The four towns of Kolding, Vejle, Fredericia and Middelfart decided as early as 1961/62 to join forces to build a common ruin town despite receiving no financial assistance from the DCDD (*CF-samvirket Gudsø* 1987). The importance and value of ruin towns seem to have been shared more generally by the local communities. When a town started building their own ruins, local newspapers celebrated the event. It was clearly a matter of great prestige and identity, as if the risk of becoming a target of the enemy's assault marked the town as particularly important. On 22 May 1966, the local paper *Vendsyssel Tidende* proudly reported that the ruin town that was being built in Hvims, Northern Jutland, 'can already demonstrate significant achievements even though it is only half-finished', and that local civil defence had 'already benefited greatly' from it.[19]

The ruin towns were all centred on the so-called standard ruin (Fig. 7.2). It was the perfect ruin. First, it was realistic in two ways: it

Fig. 7.2 DCDD blueprint of standard ruin, 1.16 Structures B and C, perspective

(Structure B imitated a relatively unharmed building with an intact ground floor, partly damaged first floor and attic. C was a heavily damaged building, where it was just possible to discern the structure of walls at ground floor; the above floors were completely damaged. In addition, there was an undamaged basement. In order to protect the interior from bad weather, a flat roof construction covered the building. The pile of rubble in front of the houses was the "remains" of a completely ruined building A. *Source* Aalborg Kommune Byggesagsarkiv C-18316 1965) looked like a real ruin, and it resembled common Danish townhouses. Second, it was safe: the training would not become too realistic as a result of actual collapse. Third, it was created with the essential civil defence tasks in mind. The standard ruin consisted of five buildings, named A–E, which represented the remains of five originally neighbouring houses in various stages of ruination.[20] (Figs. 7.3 and 7.4).

The buildings embodied different types of collapse as described in the *Basic rescue textbook* (Civilforsvarsstyrelsen 1965), fitting them for use in civil defence exercises for different types of rescue operations. The textbook did not differentiate between damage caused by conventional or nuclear bombs (Civilforsvarsstyrelsen 1965: 58). The rationale was that the main difference between the two concerned the extent of the damage and only to some degree the character of rescue tasks. The latter mainly referred to the location of potential victims, as they would not primarily be located in basements but on all storeys, because modern weapons technology (rockets and short warning times) meant that people would have little time to seek shelter (Civilforsvarsstyrelsen 1965: 58). The corresponding textbook about firefighting likewise operated with few distinctions between a conventional and a nuclear attack and drew heavily on the experiences of the Second World War (particularly the fire storms in German cities). The book noted that the nuclear attacks on Japan caused similar fire phenomena, and therefore, conflagrations would likely occur no matter what kinds of weapons were used in a future war (Civilforsvarsstyrelsen 1971: 71).

The standard ruin was built according to a fixed set of architectural blueprints developed by the DCDD. This made it possible to re-create the ruins almost identically everywhere, and the ruins became in a sense trans-spatial. In order to receive financial assistance for the construction of a ruin town, local civil defence had to use the official blueprints for the standard ruin. This ensured a standardisation of ruin building and a uniformity and homogeneity of ruins across Denmark. To make a proper ruin town, however, the standard ruin would be supplemented with other ruined buildings.

The Danish ruin towns were emulating 'X-town' (*X-købing*), a fictive but generic and conventional market town that regularly suffered attack in civil defence war games, films and planning material. Built structures typical of Danish provincial towns and common street names made the townscapes familiar and recognisable, yet not specific enough to be identifiable as one particular town. The exercises in the ruin towns were meant

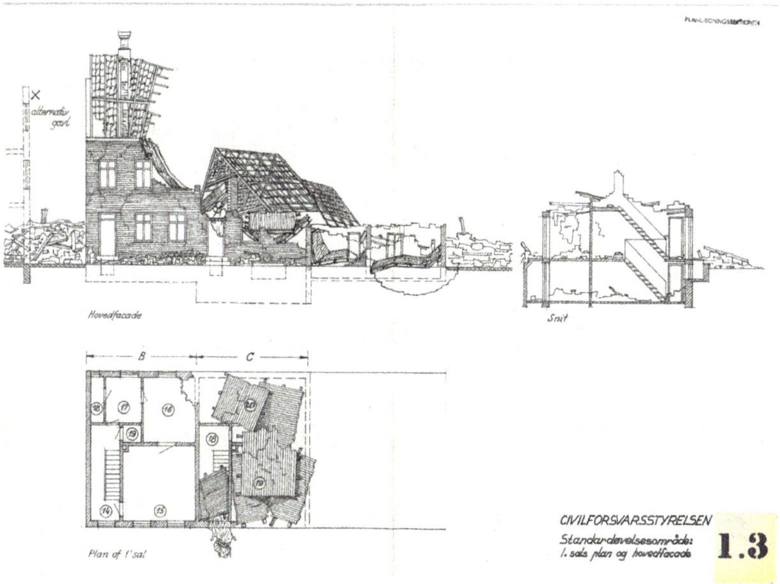

Fig. 7.3 DCDD blueprint of the standard ruin, 1.3, structures A-E, main façade and section
(The brick work was made with cracks to create an authentic look, but only on the surface as the buildings were of course not meant to actually break down. Live or mock victims would be placed in the basements and upper storeys beneath piles of debris, rubbles and crashed ceilings, from where they could be rescued, either lowered or hoisted on stretchers. Walls had to be penetrated, tunnels dug through debris, crashed floors removed and unstable walls braced to locate and rescue victims. In addition, the ruin could also serve as training ground for work in confined spaces, as well as practising the use of gas masks and cutting off gas, water and electricity. Zinc gutters were placed at windowsills in order to light fires or create smoke in the buildings (in a similar way to how it was done in Eastwood Park), but only specially designated houses, called fire houses, were actually set on fire. The standard ruin was even furnished in to make it more authentic. Chandeliers and wardrobes moreover also served as markers for orientation in dark or smoke-filled rooms. *Source* Aarhus Kommune Byggesagsarkiv Byggesag 65–79)

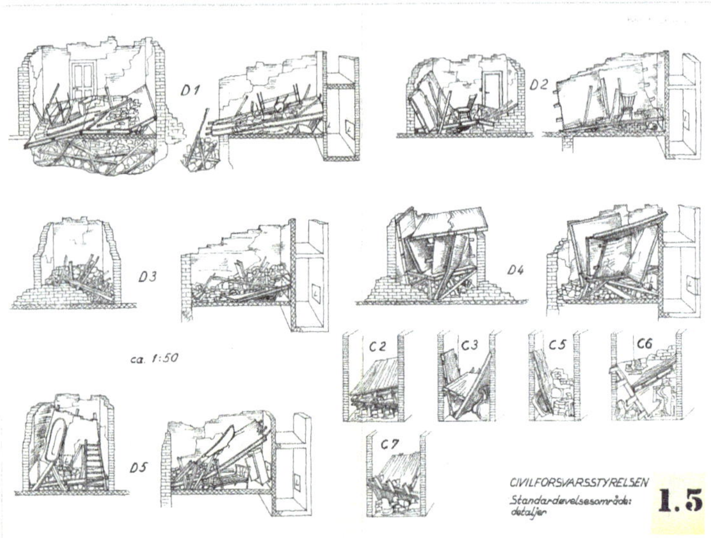

Fig. 7.4 DCDD blueprint of the standard ruin details from structures C and D
(Details from structures C and D. Detail drawings meticulously noted where and how debris, crashed ceilings and floors as well as furniture should be placed in order to ensure safety as well as facilitating different training tasks. *Source* Aarhus Kommune Byggesagsarkiv Byggesag 65–79)

to hit close to home for all civil defenders and encourage a sense of responsibility: this could be their hometown… in ruins. Leader of Hvims ruin town, J. C. Kristensen, told a local newspaper in 1966 that the ruin town had given his men a better understanding of the necessity of the training and made them more enthusiastic.[21] It confirmed the value of their work, reinforced the credibility of civil defence and strengthened the personnel's collective identity as humanitarians. In this way, the ruin towns contributed to processes of embedding the civil defence sociotechnical imaginary in local communities.

The scale and extent of ruin towns varied across the country, which may reflect different local attitudes towards civil defence, as well as different financial situations. Some ruin towns were small and modest. In Odense on the island of Funen, the standard ruin remained solitary until supplemented during the 1970s by two other buildings.[22] Other ruin towns quickly outgrew their humble origins. At Kirstinesminde, just north of Aarhus in Eastern Jutland, several streets and a dozen groups of buildings, including two fire houses and two fire ruins, were constructed. In 1976, a so-called 'corner building' was planned and built.[23] It was based on blueprints from similar buildings in Esbjerg and Aalborg, which testifies to a translocal circulation of ideas and materialities.

The corner building was a modern type of multistorey housing block built in aerated concrete. In the 1970s, this kind of built structure had begun to replace the traditional Danish brick house as the preferred building type, and it collapsed in entirely different ways, hence creating a need for updated ruins. Though this development occasioned the inclusion of a new building design, the scenarios and the training changed little.

Performing and Taming Nuclear Catastrophe

According to the instigator of Tinglev, K. M. Nielsen, the ruin town was 'a training ground where the personnel could receive education in basic rescue in as realistic conditions as possible so that no task in actual conditions would appear unfamiliar to them'.[24] The quote reveals a utopian belief that all scenarios could be predicted, planned, prepared for and performed. War, whether nuclear or conventional, could be pre-enacted—not re-enacted, as it had not yet occurred—and rehearsed in authentic environments.

The ruin town was not meant to simulate ground zero, zone A, of a bombed town, but rather zones B and C further out. Not a metropolis, nor a village, the ruin town represented an intersection between the doom of the urban ground zeros and rural safety. The building structures had suffered heavy damage, but they were not completely obliterated. There were casualties but also many survivors to be rescued. In short: catastrophic but not apocalyptic.

The tasks that were practised in the ruin towns were mainly basic rescue and first aid, firefighting and debris clearance: tasks that would be difficult—if not impossible—to carry out if an area had been contaminated by radioactivity.[25] In 1966, Tinglev started to offer tactical courses on protection against weapons of mass destruction (in Denmark called ABC weapons: atomic, biological and chemical), and in the 1970s, simulated nuclear bombs were exploded as part of the course.[26] Yet practical civil defence exercises in the ruin towns seem to have been mainly based on scenarios of conventional attacks.[27] Horrifying as these may have been, the lack of explicitly nuclear conditions is remarkable, not least given the increasing public attention to the dangers of radioactivity, particularly after the US H-bomb tests on the Bikini Atoll in 1954.

During the 1980s, there was a renewed focus on nuclear and chemical weapons in both Danish public debate and the civil defence organisation. A special Atomic Chemical Officer course was established in 1983, and the practical courses at Tinglev included instruction in how to wear protective clothing, use dosimeters and other measuring equipment, and how to cleanse after exposure to radioactivity.[28] The exercises also include scenarios of radioactive contamination.[29] However, the main focus continued to be firefighting and rescue, and the textbooks from the 1960s were still in use. The pragmatic reasoning within the organisation was that close to an urban ground zero in a thermonuclear attack, there would be nothing they could do, and thus no point in training for it. Furthermore, such an attack would leave few survivors and thus, paradoxically, less work for the civil defence (mass burials of the dead were not a civil defence task). In contrast, the conventional scenarios would demand a huge effort on the part of the civil defenders.

Following this line of reasoning, the original core activities of putting out fires and rescuing people from buildings and shelters would be similar after either a nuclear or a conventional attack, even if the circumstances were not.[30] Training for nuclear war simply added the extra dimension of being attentive to levels of radiation and getting out when these levels constituted a lethal danger to personnel, a challenge each individual civil defender was thought capable of dealing with. Whether a pragmatic and rational approach or a form of nuclear 'psychic numbing' (Lifton 1982), the attitude attuned the perception of personnel to the threat and how to live with it. Nuclear war was effectively normalised as just another war scenario that could be prepared for in the same way as non-nuclear war, and civil defence work was rationalised as functional, useful and important no matter the circumstances (see also Hogg this volume). By defining and explicating the threat, the materiality of the ruins rendered alternative scenarios invisible or insignificant, and the effect was the construction of a consensus about danger, preparedness and resilience.

The ruin town can be seen as a stage for 'performative occasions' (Jasanoff 2015a:11): training and exercises were carefully planned and orchestrated actions resulting—ideally—in familiarisation with the danger, as expressed by Nielsen. After the Cold War, international historical research often used theatre metaphors to describe civil defence (Davies

2007). This comparison was not unfamiliar to the historical actors themselves, however. Architect Grut remarked in his report following the study trip to Eastwood Park that the ruin town 'to some degree resembles a stage setting that must be constructed with a thorough sense of realism'.[31] By pre-enacting the future war in highly detailed and often publicised spectacles of theatrically rehearsed destruction in the ruin towns, the personnel were trained bodily, emotionally and mentally to instil in them experiences and habits that could be mobilised in times of crisis. Enacting the routines in the ruin town again and again can be interpreted as transforming or 'taming' (Clark 1999) an unknown, incomprehensible and incalculable catastrophe, nuclear war, into a familiar and manageable urban calamity, collapsed buildings and fires.

The ruin town offered a possibility of confronting the unknown and unfathomable catastrophe, naming it, articulating it and thus disarming it. Though scary and disturbing, the ruin towns were manageable places of destruction. They were not spaces of despair or hopelessness. Rather, they promised victory, hope and a new beginning in the postwar future. Training and exercises among the ruins demonstrated that knowledge, experience, strength and willpower (and the right technical equipment) were all that was needed to sort out chaos and re-establish the prewar social order. Far from being a radical utopian desire for a tabula rasa and an urban renewal following a nuclear apocalypse (Monteyne 2011), the ruin town can be seen as the (willed) triumph of everyday life in the provincial town, and the desire to extend existing norms and ways of life into the postwar situation.

The technoscientific imagineering of disaster potentially had social effects beyond civil defence circles. The ruin towns were used as the setting for civil defence propaganda films for a broader audience, and they featured frequently in local and national newspapers. Through these mediations, the ruin town subjected the war to public admiration (rather than scrutiny).[32] Without fail, the newspapers stressed the realism and authenticity of ruin towns and took pride in how these environments were (allegedly) capable of imitating the abysmal reality of nuclear war.[33] The notion that local civil defence forces always managed to combat the danger and rescue the (fictive) inhabitants was celebrated, for instance by the local paper *Vestkysten*, which in 1962 ran the headline: 'The efficiency of Esbjerg Civil Defence proven by realistic exercise in Tinglev'.[34] Even when a civil defence exercises actually demonstrated weaknesses or flaws, the papers often managed to turn the narrative around to something positive. In 1975, the national paper *Jyllands-Posten* reported on an exercise conducted by the national Civil Defence Corps. The paper stressed that despite severe budget cuts for years, civil defence was still capable of carrying out their tasks. There had been challenges for the exercise, however, the paper lamented, specifically the conscripts' lack of enthusiasm, which was caused by their disapproval of the war scenario. The conservative paper gave a thinly veiled critique of the indolence of the youth and their lack of understanding the seriousness of the Communist threat, which caused them to rebel. Still, the paper reassured the readers; in the end, the exercise was successful due to the untiring efforts of the officers.[35] The population could rest assured that rescue was indeed possible, even likely—if, of course, they acted rationally and behaved according to the instructions of the authorities.

The ruin town was 'politics in matter' (Weizman 2007); it staged public rituals and technoscientific practices, repetitive performances of statehood and democratic participation that formed an act of governance. The average Danish main street was cast as the target of the enemy's assault, but through rehearsing and pre-enacting, the state demonstrated, embodied and manifested its promise to take care of its citizens and re-establish social order. That particular social order was the modernistic, functional welfare state. Welfare and civil defence ideologies were alike in promoting the ideal of the rational, well-prepared citizen of modern society (in this case, the educated civil defence personnel) who contributed to the common good and survival of society by fighting and controlling the threat posed by global and total war. Performing the work in the ruin towns depended on everyone fulfilling their specific and vital task in solidarity with and for the benefit of the larger whole, which reflected a general understanding of society in Danish civil defence in stressing the social norms of conscientiousness and community spirit (Bjørnsson, Farbøl and Sylvest 2020). Acquainting civil defenders and civilians alike with the mass death that loomed in the shadow of the mushroom cloud, the ruin towns educated them in the right behaviour to avoid panic (see also Bjørnsson 2020; Sylvest 2018).

Conclusions

This chapter has attempted to demonstrate how a study of ruin towns can develop existing civil defence history by bringing an appreciation of the textured and spatialised, experiential and embodied nature of sociotechnical imaginaries. The framework of sociotechnical imaginaries attunes us to both the central role of technology in civil defence and the co-production of ideas and materialities, which is clearly exemplified by the imagineering of the ruin town. However, this chapter also demonstrates, first, an embedding-extension fuzziness and, second, that a deeper understanding of materiality is necessary to understand the multiple and complex ways that matter matters. Economic restraints impeded the realisation of the British-inspired grand design for a ruin town, yet the idea and inspiration extended easily across borders. Danish civil defence authorities were eager to import and replicate the British imaginary and the technical know-how to materialise it. Whereas extension in Jasanoff's version seems to be a result of an assertion of power by the hegemonic actor over the less powerful, whether this is desirable or not, here we are clearly dealing with what we can call 'extension by invitation' (with a rephrasing of Geir Lundestad's famous title 'Empire by invitation' [1986]).

Instead of a British village, the ruin town took the shape of a Danish provincial town; it was extended into a new national context and a new architectural tradition. Here, however, the linearity of the phases is broken, because this extension cannot be understood separately from the simultaneous embedding of a particular civil defence sociotechnical imaginary that emphasised resilience, survival and regeneration, where welfare triumphed over warfare. Through the trans-spatial structure of the standard ruin, the imaginary was then re-extended across the Danish civil defence landscape as more than 20 provincial towns all over Denmark built their own ruins of resilience during the 1960s and 1970s. Centralised state planning and design resulted in nearly identical sites and built structures with only minor variations created by local landscape, finances and devotion to the cause of civil defence. Because of the recognisability of the standard ruin and X-town, the re-extension was at the same time a re-embedding of the imaginary in local contexts, settings, minds and emotions. It was, then, no straightforward or linear process; rather, the chapter shows embedding-extension to be a two-way process that might be analysed separately but, in reality, was complex, messy and impossible to disentangle.

To substantiate Jasanoff's notions of materiality, I have sought inspiration in ruinology and New Materialism, as well as employing methods of architectural analysis in addition to traditional historical methods. Architecture for civil defence was a medium for the imagination, design and construction of spaces in preparation for a moment of ultimate crisis: nuclear war (Monteyne 2011: xv), and the ruin town telescoped the anticipated global nuclear war into the local environment, entwining materiality and imaginaries. The standard ruin in particular and the ruin town in general expose the threat perceptions of the imaginary war that shaped the organisation of Danish civil defence. While the ruins reflected civil defence's preparation for what they were capable of acting on, the material structures themselves also became agents of continuity. The war-torn town in brick and concrete fostered a form of path dependency that contributed to an inertia and sturdiness in the perception of the future war and how it could be mitigated.

The ruin town's main raison d'être was the (imitation of) realism and authenticity, which made it possible to rehearse and pre-enact the imagined future war. The civil defence exercises that took place among the ruins attempted to foster a special—and spatial—idea of a Third World War balancing between utopian salvation and dystopian obliteration. Unlike real ruins in war-torn cities, the ruin towns were predictable, predesigned and predetermined, and they standardised an unknowable future. They made it possible to assert control over the thought and behaviour of the users and spectators by conforming to coercive programmes of use and preestablished ideologies to enforce a unity of meaning and materiality (Wood 1993), but they also structured and set the limits of imagination, which testifies to the solidity as well as agency of things.

In effect, the ruin town contributed to a taming of the unknown and unknowable catastrophe. It demonstrates the 'blurred the lines between real and imagined realities' (Jasanoff 2015a: 5); the 'real' reality of insurmountable destruction by thermonuclear warfare was tamed through a collective desire for logic and order to produce an imagined reality of a manageable urban calamity and an authoritative but normative representation of how the war ought to and would play out—and namely, how it was eventually managed and overcome.

The ruin town gave shape, mass and solidity to the imagined nuclear catastrophe, to the horror, danger, darkness, agony and fatalities associated with it. Yet it did not present the end of the world as it was known. Far from an 'architecture of disillusion' (Vidler 2010), the ruins, rather, suggested the resilience and eventual triumph of the welfare state and existing social order and norms in a postwar world. It thus reflected and reinforced a specific, historically situated sociotechnical imaginary in which the 'good society' was protected from the ultimate social, political and environmental chaos, the nuclear war.

Acknowledgements The author wishes to extend her sincere gratitude to the personnel at the Danish Emergency Management Agency and the municipal Emergency Management organisations who have provided invaluable assistance and gratuitously taken her on tours through the ruin towns of Denmark. In particular, she would like to thank Eva-Marie Nielsen (Tinglev), Thorbjørn Jørgensen (Kirstinesminde), Daniel Birkebæk (Odense), Basse Vængtoft (Hedehusene) and Jan Nedza and Søren Korsgaard (Rørdal).

Notes

1. RA, ES 132 Diverse rapporter og betænkninger, 3L 'Rapport over undertegnede og arkitekt Gruts rejse til England…', 1951.
2. *Vestjyden* '"Civilforsvarets leder hjem fra England efter at have studeret instruktørskole', 21 April 1951, 6.
3. This literature is focused on 'real' ruins: buildings that have become ruins as a result of neglect, accident or war. Still, there are a number of insights, understandings and ideas that are fruitfully harnessed also to a study of artificial ruins.
4. RA, ES 132/3.
5. RA, ES 132/3 'Rapport over…', 1951.
6. RA, ES 132 'Tillæg', 1951.
7. This part of the chapter is primarily based on unarchived sources from the technical school in Tinglev kindly placed at the author's disposal by section leader Eva Marie Nielsen in February 2020: BRS Tinglev, files 'Ruinanlæg', Notes about the ruin town 'Ruinbyen i Tinglev' by K. M. Nielsen, dated December 1955; BRS Tinglev, files 'Diverse' M. Hedges, 'Denmark's system of mobile columns', *The Municipal Journal*, 25 September 1959; *Vestkysten* 28 June 1968, 'Fra krigshjælpsarbejde til opgaver i fredstid', 7; *Ekstra-Bladet*, 21 December 1961 'Han byggede en ruinby'. Nielsen was key to the construction of the ruin town, which was described by the later director general of the DCDD as basically a 'one man job', Rigsarkivet (henceforth RA), Beredskabsstyrelsen

(henceforth BS), Landsarkivet for Sønderjylland (henceforth LS) ME/-879 Teknisk skole, Skolebefalinger, April 1961.

8. BRS Tinglev, 'Ruinbyen i Tinglev', 1955.
9. See blueprint in BRS Tinglev, files 'Ruinanlæg' as well as description in 'Ruinbyen i Tinglev', 1955.
10. BRS Tinglev'Ruinbyen i Tinglev', 1955; *Horsens Social-Demokrat*, 13 January 1959 'En uhyggelig ruinby, der 'kun' kostede 20.000 kr.', 3.
11. RA ES Breve 62/Breve 1957–1958 Letter from Schultz to Abelin dated 11 February 1958. Interestingly, Schultz sent Grut's original blueprints to his Swedish counterpart, but whether the plans were realised in Sweden is not clear.
12. 13 January 1959, p. 3. Falfield had 22 buildings according to Petersen's report, compared to Tinglev's six. The article also exaggerated in claiming the price of Falfield was DKK 2 million and Tinglev just DKK 10,000. See also *Jydske Tidende (JT)* 19 November 1955 'Ruinby i København efter tegninger fra Tinglev', 9.
13. *JP* 9 March 1957 'Civilforsvaret får nu en teknisk skole i Tinglev', 6.
14. *Berlingske Tidende*, 29 January 1962 'De lærer civilforsvar i sønderjysk ruinby', 8; *Herning Folkeblad* (*HF*), 2 October 1961 'Fra Hamborg til Tinglev for at studere – ruiner', 3; *Horsens Social-Demokrat*, 13 January 1959; *JP*, 9 March 1957; *JT*, 19November 1955; *Vestkysten*, 9 February 1962 'Hele Esbjergs civilforsvar til stor øvelse i 'Ruinbyen', 14; *Vestkysten*, 16 September 1969 'Tinglevs store ruinby giver inspiration til tre norske CF-ledere',17; *Vestkysten*, 21 January 1975 'Arabisk CF-general på besøg i Tinglev', 2; *JT*, 25 May 1968 'Nordisk kontakt i civilforsvaret'; *JT*, 31 January 1975 'Saudi-arabere lærer civilforsvar'; *JT*, 28 October 1966 'Thailændere på CF-skolen'.
15. BRS Tinglev, file Ruinanlæg, blueprint and map from 1958; Rigsarkivet (henceforth RA), Beredskabsstyrelsen (henceforth BS), Landsarkivet for Sønderjylland (henceforth LS) ME/-879 Teknisk skole, Skolebefalinger, January 1960, April–May 1961.
16. *Civilforsvarsbladet* 1963:10. Seks øvelsesanlæg sættes i gang, 15.
17. There were four facilities in CD region I (Northern Jutland), including Rørdal, Hobro and Hvims (a cooperation between the towns Skagen, Frederikshavn, Sæby, Hjørring and Hirtshals);

seven in region II (mid-Jutland), incl. Holstebro, Randers, Århus, Viborg, Herning; four in region III (Southern Jutland), incl. Gudsø (cooperation between Vejle, Fredericia, Kolding and Middelfart), Horsens, Esbjerg; three in region IV (Funen and islands), Assens, Odense, Svendborg; four in region V (South-Western Sealand), incl. Næstved and Nykøbing F.; three in region VI (Northern Sealand and Copenhagen), Helsingør, Køge and Birkerød, later Hedehusene; and finally one in region VII (Bornholm), Rønne. In addition, there were a number of 'mini training facilities'. RA, Årsrapport 1972 and 1977.

18. Molitor notes how West German civil defence thrived in détente periods, not during crises, see Chapter 3 this volume.
19. *Vendsyssel Tidende*, 22 May 1966 'Byen i Hvims, hvor man bygger vakkelvorne huse', 15.
20. This description of the standard ruin and the blueprints are all located at Aarhus Municipality's online archive of construction projects, 'Min ejendom' https://minejendom.aarhus.dk/Byggesag/Liste?adresseId=58345 [accessed 10 February 2020], see specifically Byggesag 65–79 'Entreprisebeskrivelser' and 'Etageplaner'.
21. *Vendsyssel Tidende*, 22 May 1966 'Byen i Hvims, hvor man bygger vakkelvorne huse', 15.
22. Odense, Åsumvej 55, Ejby, Odense Jorder, 1h, journalsag 577 Standardøvelsesanlæg, 1159 'Flyt. Undervisningsb.', 1055 'Tilbygning' https://www.weblager.dk/app [accessed 10 February 2020].
23. AS, Århus Kommune Byrådet Journalsager 1985, 'Projekt 1976'. See also Aarhus Municipality's online archive of construction projects https://minejendom.aarhus.dk/Byggesag/Liste?adresseId=58345 [accessed 10 February 2020].
24. BRS Tinglev 'Ruinbyen i Tinglev', 1955.
25. A similar bypassing of the dangers of radioactivity is noted by Hogg in his chapter on civil defence exercises in Britain, see Hogg this volume.
26. *Vestkysten* 10/6/1970 'Atombombe' sprænges på tirsdag ved Søgård'; *JT* 16 June 1970 'A-bombe' øvelse'; *JT* 10 August 1971 'CF sprænger mini-atombombe'.
27. Unfortunately, civil defence exercise scenarios have not been saved for posterity at Tinglev or any other of the ruin town that has been

selected for field studies; hence, the primary descriptions of these exercises still in existence are brief newspaper reports.

28. *Vestkysten*, 17–18 December 1983 ''Atom-truslen sætter helt ny uddannelse i gang på CF-skole'.
29. *JT*, 20 April 1986 'Da bomberne ramte Tinglev'; Personal conversation with former CD officers H. A. J. Larsen, Barak H1 Beredskabsforbundets informations- og udstillingsbarak, 12 February 2020 and Thorbjørn Jørgensen, Civil Emergency Preparedness Inspector at Aarhus Beredskabscenter, 13 February 2020.
30. Personal conversation with H. A. J. Larsen 12 February 2020 and Thorbjørn Jørgensen 13 February 2020.
31. RA, ES 132 'Tillæg', 1951.
32. As Cronqvist and Grant (this volume) remind us, individuals experienced civil defence in numerous different ways, and it is beyond the scope of this chapter to evaluate the degree of success of extending a specific sociotechnical imaginary to a wider population through the ruin towns. The reassuring message of resilience and survival was certainly picked up by national and local newspapers, but whether the population accepted and internalised this sociotechnical imaginary remains a question for future research.
33. *Berlingske Tidende*, 29 January 1962; *Bornholms Tidende*, 1 December 1969 'CF-øvelse årsag til blind brandalarm', 4; *Fredericia Dagblad*, 23 January 1960 'CF-ruinbyen endnu kun en ønskedrøm blandt lederne', 3; *Fynsk Aktuelt*, 26 October 1965, 'Arkitekttegnet ruin kostede 50.000 kr.', 6; *HF* 2 October 1961; *HF*, 23 October 1976 'Hård brandtørn for genindkaldte CFere', 3; *Horsens Social-Demokrat*, 13 January 1959, 3; *JP* 9 March 1957; *JP*, 20 April 1959 'Atombombeøvelse i Ruinbyen i Tinglev', 3; *Middelfart Venstreblad*, 26 October 1965 'Mange mænd træner mange timer for a t redde – ja måske Dem', 4; *Skagen Avi,s* 14 January 1961; *Vendsyssel Tidende*, 22 May 1966; *Vestkysten*, 9 February 1962; *Vestkysten*, 9/4/1962 'Esbjerg CFs effektivitet bevist ved realistisk øvelse i Tinglev', 8; *Vestkysten,* 28 June1968; *Aalborg Stiftstidende*, 19 February 1965 'Bevilling givet til ruinhus i Rørdal', 1–2; *JP*, 20 April 1959.
34. *Vestkysten*, 9 April 1962 'Esbjerg CFs effektivitet bevist ved realistisk øvelse i Tinglev', 8.
35. *JP*, 24 February 1975 'CF-værnepligtige viste manglende rutine på øvelse', 2.

References

Unpublished Sources

BRS Tinglev: Beredskabsstyrelsen Tinglev [the Danish Emergency Management's Technical School Tinglev].

1. BRS Tinglev, files 'Ruinanlæg'.
2. BRS Tinglev, files 'Standard øvelsesanlæg'.
3. BRS Tinglev, files 'Diverse'.

RA: Rigsarkivet [the National Archives] Beredskabsstyrelsen.

4. RA, ES Erik Schultz Embedsarkiv, 62 Breve 1957–58.
5. RA, ES 132 Diverse rapporterog betænkninger.
6. RA ES 231 Årsrapporter.

RA, LS: Landsarkivet for Sønderjylland [Regional archives of Southern Jutland] Beredskabsstyrelsens tekniske skole [Danish Emergency Management Agency's technical school], ME/-879, skolebefalinger [school orders].

AS: Aarhus Stadsarkiv [Aarhus City Archives],

7. AS ÅrhusKommune Byrådet Journalsager 1985, 1692 'Projekt 1976'.

Online Sources

https://minejendom.aarhus.dk/Byggesag/Liste?adresseId=58345 [10/2/2020] Aarhus Municipality's online archive of construction projects

https://www.weblager.dk/ [10/2/2020] Online archive of municipal construction projects in Denmark.

Newspapers

Aalborg Stiftstidende.
Berlingske Tidende.
Bornholms Tidende.
Ekstra-Bladet.
Fredericia Dagblad.
Fynsk Aktuelt.
Herning Folkeblad (HF).
Horsens Social-Demokrat.
Jyllands-Posten (JP).
Jydske Tidende (JT).
Middelfart Venstreblad.
Roskilde Dagblad.
Vejle Amts Folkeblad.
Vendsyssel Tidende.

Vestjyden Esbjerg.
Vestkysten.
Skagen Avis.
Sønderjyden.

Literature

Bennett, Tony and Patrick Joyce, eds. 2010. *Material powers: Cultural studies, history and the material turn.* London and New York: Routledge.

Beredskabsstyrelsen. 2006. *Beredskabsstyrelsens Tekniske Skole i 50 år: fra folkehøjskole til beredskabsskole.* Tinglev: Beredskabsstyrelsen, Teknisk skole.

Bishop, R. and G. Clancey. 2004. The City-as-Target, or Perpetuation and Death. In *Cities, war, and terrorism: Towards an urban geopolitics*, ed. Graham Stephen, 54–75. Oxford: Blackwell.

Bjørnsson, Iben. 2020. 'Stands tilløb til panik'. Civilforsvarspjecer som social kontrol. In *Atomangst og civilt beredskab: Forestillinger om atomkrig i Danmark 1945–1975*, eds. Marianne Rostgaard and Morten Pedersen, 65–102. Aalborg: Aalborg Universitetsforlag.

Bjørnsson, Iben, Rosanna Farbøl and Casper Sylvest. 2020. Hvis krigen kommer: Forestillinger om fremtiden, *Kulturstudier* 11: 33–61.

CF-samvirket Gudsø i 25 år (1987).

Civilforsvarsbladet 1963:10. Seks øvelsesanlæg sættes i gang, 15

Civilforsvarsbladet 1974:2. Poeten og lillemor i CF's tjeneste, 11.

Civilforsvarsstyrelsen. 1965. *Lærebog i redningstjeneste.* København: J.H. Schultz Universitets- Bogtrykkeri.

Civilforsvarsstyrelsen. 1971. *Lærebog i brandtjeneste.* København: J.H. Schultz Universitets- Bogtrykkeri.

Clarke, Lee. 1999. *Mission improbable: Using fantasy documents to tame disaster.* Chicago: University of Chicago Press.

Cordle, Dan. 2012. Protect/Protest: British nuclear fiction of the 1980s. *British Journal for the History of Science* 45: 653–69.

Dahl, Arthur. 1955. Countering vulnerability. *Danish Foreign Office Journal* 1: 16–18.

Davis, Tracy 2007. *Stages of emergency: Cold War nuclear civil defense.* Durham: Duke University Press.

Derrida, Jacques. 1984. No apocalypse, not now (Full speed ahead, seven missiles, seven missives). *Diacritics* 14: 20–31.

Deville, Joe, Michael Guggenheim and Zuzana Hrdličková. 2014. Concrete governmentality: shelters and the transformations of preparedness. *The Sociological Review* 62(S1): 183–210.

Edensor, Tim. 2005. *Industrial ruins. Space, aesthetics and materiality.* New York: Berg.

Eshel, Amir. 2010. Layered time: Ruins as shattered past, ruins as hope in Israeli and German landscapes and literature. In *Ruins of modernity*, eds. Julia Hell and Andreas Schörle, 133–151. Durham and London: Duke University Press.
Farbøl, Rosanna. 2020. Urban civil defence: Imagining, constructing and performing nuclear war in Aarhus. *Urban History*, 1–23https://doi.org/10.1017/S0963926820000590
Frohman, Larry. 2011. Tony Bennett and Patrick Joyce (eds.) Material powers. Cultural studies, history and the material turn. *Social History* 36: 378–380.
Grant, Matthew. 2010. *After the Bomb: civil defence and nuclear war in Britain, 1945–68*. London: Palgrave Macmillan
Hell, Julia and Andreas Schörle. 2010. Introduction. In *Ruins of modernity*, eds. Julia Hell and Andreas Schörle, 1–17. Durham and London: Duke University Press.
Hogg, Jonathan. 2012. 'The family that feared tomorrow': British nuclear culture and individual experience in the late 1950s, *British Journal for the History of Science* 45: 535–549.
Huyssen, Andreas. 2010. Authentic ruins: Products of modernity. In *Ruins of modernity*, eds. Julia Hell and Andreas Schörle, 29–41. Durham and London: Duke University Press.
Ireland, Tracy and Jane Lydon. 2016. Rethinking materiality, memory and identity. *Public History Review* 23: 1–8.
Jasanoff, Sheila. 2015a. Future imperfect: Science, technology and the imaginations of modernity. In *Dreamscapes of modernity*, eds. Sheila Jasanoff and Sang-Huyn Kim, 1–33. Chicago: Chicago University Press.
Jasanoff, Sheila. 2015b. Imagined and invented worlds. In *Dreamscapes of modernity*, eds. Sheila Jasanoff and Sang-Huyn Kim, 321–341. Chicago: Chicago University Press.
Jasanoff, Sheila and Sang-Huyn Kim, eds. 2015. *Dreamscapes of modernity: Sociotechnical imaginaries and the fabrication of power*. Chicago: Chicago University Press.
Lifton, Robert Jay. 1982. Beyond psychic numbing: a call to awareness. *American Journal of Orthopsychiatry* 52: 619–629.
Masco, Joseph. 2005/2006. Fantastic city: Engineering Ruins in Cold War America. *Cabinet* 20: 1–6.
Masco, Joseph. 2008. 'Survival is Your Business': Engineering ruins and affect in nuclear America. *Cultural Anthropology* 23: 361–398.
Monteyne, David. 2011. *Fallout shelter: Designing forcivil defense*. Minneapolis, MN: University of Minnesota Press.
Monteyne, David. 2004. Shelter from the elements: Architecture and civil defense in the early Cold War. *The Philosophical Forum* 35: 179–199.

Puff, Helmut. 2010. Ruins as models: Displaying destruction in postwar Germany. In *Ruins of modernity*, eds. Julia Hell and Andreas Schörle, 253–70. Durham and London: Duke University Press.

Stafford, James. 2012. 'Stay at home': The politics of nuclear civil defence, 1968–83. *Twentieth Century British History* 23: 383–407.

Sylvest, Casper. 2018. Atomfrygten og civilforsvaret. *temp - tidsskrift for historie* 16: 16–39.

Sylvest, Casper. 2021. Pre-enacting the next war: The visual culture of Danish civil defence in the early nuclear age. *Cold War History*, 1–23. https://doi.org/10.1080/14682745.2021.1874932

Vidler, Anthony. 2010. Air war and architecture. In *Ruins of modernity*, eds. Julia Hell and Andreas Schörle, 29–41. Durham and London: Duke University Press.

Weizman, Eyal. 2007. *Hollow land: Israel's architecture of occupation*. London: Verso.

Weizman, Eyal. 2011. *The least of two possible evils: Humanitarian violence from Arendt to Gaza*. London: Verso.

Weizman, Eyal. (2012) *Forensic architecture: Notes from fields and forums*. Ostfildern: Hatje Cantz.

Wood, Lebbeus. 1993. *War and architecture: Rat i arhitektura*, Pamphlet Architecture 15. Princeton: Princeton Architectural Press.

CHAPTER 8

Framing Civil Defence Critique: Swiss Physicians' Resistance to the Coordinated Medical Services in the 1980s

Sibylle Marti

The threat of a future nuclear war bothered Switzerland's national defence planners from the time of the early Cold War. Military, civil defence and administrative officials alike reckoned that the use of nuclear weapons would cause far more civilian than military casualties, and the results of a large national defence exercise in 1963 confirmed that the existing measures to protect and rescue the civilian population from weapons of mass destruction were not sufficient at all (Marti 2015: 252–254). That is why from the mid-1960s, Swiss federal authorities not only worked on an ambitious and world-leading shelter programme (Berger Ziauddin 2017), but also set about establishing various alarm organisations and protective services as essential pillars of Switzerland's national defence. Among these evolving national defence organisations were the so-called

Present Address:
S. Marti (✉)
Historical Institute, University of Bern, Bern, Switzerland
e-mail: sibylle.marti@hist.unibe.ch

M. Cronqvist et al. (eds.), *Cold War Civil Defence in Western Europe*,
https://doi.org/10.1007/978-3-030-84281-9_8

Coordinated Medical Services as a joint venture of military, civil defence and private organisations. Based on an all-encompassing system of underground hospitals and first-aid stations, the Coordinated Medical Services were intended to 'improve the chances of survival of the entire Swiss people (*Schweizervolk*)' in a future nuclear war (see also Chapter 3 on disaster medicine in West Germany by Jochen Molitor in this volume).[1]

It took almost two decades to complete the basic conceptual, organisational and legal work in order to establish the Coordinated Medical Services (Marti 2020: 399–408). Hence, the process of finally embedding the Coordinated Medical Services into Switzerland's national defence system coincided with renewed Cold War tensions and nuclear rearmament on both sides of the 'Iron Curtain' at the beginning of the 1980s. On the one hand, this era of the Cold War, often referred to as the 'Second Cold War', was dominated by re-emerging nuclear threats and nuclear fear (Conze et al. 2017; Becker-Schaum et al. 2016). On the other hand, a transnational peace movement (Zieman 2009; Wittner 2003) and global organisations of concerned scientists like the International Physicians for the Prevention of Nuclear War (IPPNW) (Kemper 2016) voiced disarmament and peace demands resonating more than ever in the public sphere. Furthermore, both international and national aftermath studies showed that no healthcare system would be able to cope with all the injured and burned people after a nuclear war (Berger Ziauddin and Marti 2020).

In Switzerland, as well as in many other countries, manifold forms of protest against the danger of nuclear war took place. In the early 1980s, tens of thousands protested for peace and against nuclear armament (Bein and Epple 1986; Epple 1986). Besides transnationally voiced calls for nuclear disarmament and peace, the Swiss people, as part of the protest, also raised their voices against the dominance and ubiquity of the militia army and the system of civil defence, with its countless protective bunkers. Above all, criticism was directed against the national defence officials' optimistic assumption that a nuclear war was survivable, despite its absolutely devastating short- and long-term effects (Meier and Meier 2010: 224–226). Moreover, during the course of the 1980s, a number of popular referendums directed against the armed forces and national defence, respectively, took place. This opposition culminated in a national referendum on the abolishment of Switzerland's militia army in 1989 (Degen 2017). Due to these protests in the realm of civil society, in the last decade of the Cold War, the belief in a strong national defence showed clear signs of erosion.

With regard to growing civil defence criticism, the Swiss branch of the Physicians for Social Responsibility (PSR) and the IPPNW—PSR/IPPNW Switzerland, founded in 1981—played a major role. PSR/IPPNW Switzerland also became a key actor in protest against the Coordinated Medical Services. The cases of physicians and members of PSR/IPPNW Switzerland who acted as conscientious objectors by refusing to serve in the Coordinated Medical Services as part of their mandatory civil defence service gained broad public attention. In fact, conscientious objection of service in both the armed and the civil defence forces formed a vital part of Swiss Cold War history of the 1970s and 1980s that so far has only received little attention (Schweizer 2017; Koller 2008; Möcklin 1998; Winet 1991). Another form of protest by PSR/IPPNW Switzerland was a public campaign against the newly introduced and highly contested conscription of, in particular, female health workers into the Coordinated Medical Services. PSR/IPPNW Switzerland's resistance bothered the authorities and brought Swiss state security, among others, onto the scene.

The Coordinated Medical Services were an essential organisation in Cold War Switzerland's national defence system, usually referred to as total national defence. Its main pillars were the military and the civil national defence, the latter including civil defence, state security, wartime economy and foreign policy (Degen 2009). The ideological base rested on the so-called *Geistige Landesverteidigung*, roughly translatable as spiritual or intellectual defence. It was meant to serve as an integrative and identity-forming belief system which summoned national symbols and myths like Switzerland's militia-style army and the centuries-long determination to defend home ground (Tanner 1999).

In this chapter, I will argue that Swiss authorities propagated a sociotechnical imaginary (see Chapter 1, this volume) through the Coordinated Medical Services and other total national defence organisations. The core of this total national defence imaginary was that Switzerland would be able to survive a nuclear war through a huge collective effort. In Cold War Switzerland, this vision of total national defence was quite strong, and it could exert normative authority, as it was borne and institutionalised by various state and private organisations on all federal levels. What measures and institutions were the Coordinated Medical Services built on, and which material and social resources did they try to mobilise?

How should the Swiss total national defence imaginary wield its power? By contrast, under these circumstances, how was PSR/IPPNW Switzerland able to carry out protest actions against the Coordinated Medical Services? Which discursive strategies and practices did PSR/IPPNW Switzerland use to articulate resistive behaviour, and which alternative vision was thereby disseminated?

In what follows, in the first part of my chapter, I will elucidate the system of the Coordinated Medical Services as an emblematic manifestation of the sociotechnical imaginary of Swiss total national defence. By focusing on PSR/IPPNW Switzerland, the second part of my chapter will examine the resistance against the Coordinated Medical Services. In the broader context of a newly evolving, transnational civil defence criticism in the last decade of the Cold War, PSR/IPPNW Switzerland's protest actions brought up novel, subjectivised forms of resistance to Swiss total national defence that played a vital role in contributing to the dissolution of the respective imaginary. In the third part of my chapter, I will show how Swiss authorities addressed and handled the criticism of nuclear civil defence spread by PSR/IPPNW Switzerland. I will also point out how the latter's protest actions, as well as its counter-vision to the total national defence imaginary, resonated with the Swiss public. The conclusion will highlight that PSR/IPPNW Switzerland's subjectivised forms of resistance were a strategic reaction to the contested vision of Swiss total national defence in the 1980s.

Mobilising the Nation to Save the 'Swiss People'

The Coordinated Medical Services relied on an extensive infrastructure, most of it built underground. In its final expansion, the system would be equipped with almost 2,000 first-aid stations and 30 emergency hospitals maintained by civil defence staff. Another 150 protected operating rooms would be operated by public health organisations. Finally, there would be 40 military hospitals run by army personnel. In sum, these facilities were to provide protected patient spaces for two per cent of the population spread all over Switzerland.[2] At the beginning of the 1980s, about half of these facilities were ready to work. Until all of them were built, the Coordinated Medical Services also had to use above-ground first-aid and hospital rooms (Senn 1983: 84–85).

For this ambitious system to be implemented, the Coordinated Medical Services needed the support of state and private actors from all federal levels. On the national level among others, the Federal Military Department, the Federal Office for Civil Protection and the Federal Office of Public Health were involved. Furthermore, representatives of the Swiss cantons and of several private organisations like the Swiss Red Cross, the Swiss Samaritan Foundation, the Swiss Medical Association and the Swiss Hospital Association formed part of the establishment of the Coordinated Medical Services.[3] In order to guarantee the smooth cooperation of these different actors, coordination bodies on both the national and cantonal levels were installed (Senn 1983: 85).

The leading threat for the creation of the Coordinated Medical Services was the danger of a nuclear war. Right from the beginning, however, the authorities planned the Coordinated Medical Services to also be deployed in the case of a civil (nuclear) catastrophe.[4] Within the Coordinated Medical Services, the civil authorities, the civil defence organisation and the armed forces were responsible for running and maintaining their own sanitary facilities (Senn 1983: 85). Thus, the Coordinated Medical Services were an emblematic Swiss total national defence organisation for which it was characteristic that neither military nor civil institutions took the lead. Prevailing military and civil tasks and structures were kept, and the organisation was to be used in mass catastrophes taking place in war as well as in peacetime. That is why one cannot speak of a centralisation of the Coordinated Medical Services nor, as some critics have argued, of a militarisation of Switzerland's public health system (Lauterburg 1988: 215). Rather, the Coordinated Medical Services should be understood as a hybrid organisation which blurred the boundaries between the states of war and peace. Exactly because of this hybrid structure, Switzerland's total national defence organisations were able to permeate all parts of society (Marti 2020: 417–418).

The Coordinated Medical Services' all-encompassing system required not only considerable infrastructural equipment and material supply but also a huge number of human resources. Besides the members of the army and civil defence medical services, the Coordinated Medical Services needed a great number of voluntary staff. Because this voluntary recruitment could not fill the ranks, Swiss authorities also started to conscript health workers (such as physicians, nurses, physiotherapists). With these efforts to introduce conscription for the Coordinated Medical Services, the authorities tried to install mandatory service for women in particular. Unlike Swiss men, Swiss women were not obligated under the Swiss federal constitution to serve (Bondolfi 2012; Schindler 1997), but they represented the majority of employees in Switzerland's health system (Brändli 2012). With the general conscription for men, it was fairly easy to involve male medical and nursing staff in the Coordinated Medical Services, whereas women had to volunteer for the Swiss Red Cross or the civil defence medical services.

Thus, on the federal level, there was no legal basis to make women serve in Switzerland's total national defence system. Furthermore, in the realm of public health, the cantons and not the federal government had the legal power. That is why the concept of the Coordinated Medical Services envisaged the cantons as issuing legal regulations in order to conscript women (Lauterburg 1988: 208–211). To that end, the concept of the Coordinated Medical Services provided a template for the cantons to use.[5] At the end of the 1980s, most cantons had implemented such regulations or were just about to do so. With this approach, the cantons operated in a legal grey zone. That is why a contemporary critic called it a 'women's compulsory service in a roundabout way' (Lauterburg 1988: 208). But the Federal Office of Justice also acknowledged that this legal opinion must be sanctioned by the Federal Supreme Court of Switzerland (Lauterburg 1988: 208–211). In the decision in a leading case, the Federal Supreme Court of Switzerland in May 1989 sustained the cantonal authorities' legal measures. In their grounds for the judgement, it specifically pointed out that the civil defence organisation, the Red Cross service and the women's military service suffered from staffing shortages of tens of thousands of women (Bundesgerichtsentscheid 1989). Thus, the Federal Supreme Court of Switzerland ruled that the questionable conscription of female nursing and medical staff for the Coordinated Medical Services which the cantons had installed was constitutional. It is telling that this legal practice was sanctioned with reference to the requirements of Switzerland's total national defence.

Apart from these legal options, the authorities undertook a variety of information, training and advertising efforts to recruit more staff. These efforts aimed at reaching retired health workers and, first and foremost again, women. At the very beginning, when the Coordinated Medical Services were yet to be established, the surgeon general of the Swiss armed forces, who was the coordinating leader of the Coordinated Medical Services at the federal level, did much of the public relations work. He held lectures for civil defence units and officers' societies, as well as at congresses and training courses, in which he pointed out the necessity and urgency of creating the Coordinated Medical Services. In addition, he wrote a great number of articles for military and other journals.[6] Later on, standardised lectures and an official information concept were formulated.[7] The information concept was intended help to promote a 'common line of thought, a *Unité de doctrine*' between the different actors involved in the Coordinated Medical Services, and, through image advertising and the fostering of community spirit, 'enhance the "Good Will"' for this total national defence organisation.[8]

With regard to training efforts, the Coordinated Medical Services worked together with several state and private organisations such as medical schools and organisations engaged in the education of healthcare personnel. Upon an initiative of the surgeon general of the Swiss armed forces, the medical schools integrated the discipline of disaster medicine into their curricula.[9] The healthcare personnel also received training in the field of disaster and war medicine. The students at the Swiss Red Cross nursing schools, for example, had to take lessons about 'nursing in disaster situations'.[10] Other Red Cross courses designed for lay people likewise included teaching units to serve in the Coordinated Medical Services (Lauterburg 1988: 209). The link between training and public relations efforts that was pushed with these courses was deliberately planned. Training and promotion for the Coordinated Medical Services were to go hand in hand and be designed, in particular, to recruit women. The surgeon general of the Swiss armed forces, for example, urged members of the Swiss Federation for Civil Protection and the Swiss Officers Society to 'win over our women *for service for the community* everywhere where you have influence'.[11] At the same time, advertising campaigns promoted 'the woman in service of society as a whole'.[12] Accordingly, as a counterpart to the mandatory military or civil defence service of Swiss men, Swiss women were called upon to carry out their patriotic duty for the nation by serving in the Coordinated Medical Services. Playing on a socially embedded association between women and social care, it was thought they would take on nursing and auxiliary tasks, which were construed as being female in nature. Thus, Swiss total national defence measures and institutions were deeply shaped by traditional gender roles.

All in all, the Coordinated Medical Services' massive infrastructure reflected the vision and the social order of Switzerland's total national defence system, which both, in turn, were expected to facilitate the Coordinated Medical Services' required material supply and human resources.

In fact, many different societal groups and institutions, such as members of the armed and the civil defence forces, cantonal and communal authorities, organisations like the Swiss Red Cross or the Swiss Medical Association and public and private hospitals, worked together in order to establish the Coordinated Medical Services. In doing this, they all helped to promote the sociotechnical imaginary of total national defence and its core conviction that Switzerland as a collective was capable of surviving nuclear war. This imaginary materialised in the Coordinated Medical Services' all-encompassing infrastructure and was institutionalised in the organisation's hybrid structure. At the same time, state and private, military and civil actors of all federal levels alike were called to contribute to and participate in the organisation. An identical call was directed towards Swiss women. Thus, the Swiss total national defence imaginary that was propagated through the Coordinated Medical Services relied constitutively on the mobilisation of society as a whole.

But the numerous efforts to mobilise the nation in order to save the nation were only partly successful. Above all, the number of the much-needed women could not close the estimated staffing gaps. That is why the chief of general staff of the Swiss armed forces was still complaining about a staffing shortage of 75,000 female volunteers in 1983 (Senn 1983: 85). The reason for this, on the one hand, was that it proved to be fairly complicated to record all the (female) medical and nursing staff and trained lay people that could have been forced to serve by law (Lauterburg 1988: 210). On the other hand, as the next section of my paper will outline, the growing criticism directed against the Swiss armed and civil defence forces in the 1980s further hindered the recruitment of (female) volunteers.

Practices of Resistance: Claiming That Individual Action Matters

In the course of the 1980s, despite being prosecuted, more and more Swiss men refused to serve in the civil defence forces. Although the actual number was officially unknown, politicians and civil defence officials alike stated in the media that the number of conscientious objectors seemed to be increasing steadily. Sometimes, a group of people collectively exhibited their refusal to serve. In August 1984, in Bern, for example, 17 men jointly returned their documents to the city's civil defence office. Likewise, in November 1984, in Zurich, 38 men collectively embedded their service books (*Dienstbüchlein*) for the civil defence service in concrete and dumped them all in a river.[13] Against this backdrop, the cases of members of PSR/IPPNW Switzerland who did not want to serve in the Coordinated Medical Services as trained physicians gained widespread public attention. This all the more so as their refusal led to official charges and, as a result, to media-effective trials (Albrecht et al. 1988: 12).

A textbook case, in this respect, was Thomas Schnyder, a child psychiatrist and board member of PSR/IPPNW Switzerland, who during the 1980s faced several official charges due to his refusal to serve in the civil defence forces. Before 1979, Schnyder had served in the Swiss armed forces and from March 1979 to February 1983 in the civil defence forces.[14] From November 1983 onwards, however, he no longer attended mandatory civil defence courses and exercises. In May 1984, he asked the civil defence office of the canton of Zurich to only assign him to services related to natural or civil disasters, like providing surgical assistance in a hospital, and to release him from any services related to nuclear war and nuclear civil defence.[15] The civil defence office of the canton of Zurich denied this demand.[16] In April 1985, Schnyder sent the same request to the municipal civil defence office in his residential community.[17] This request remained without reply.[18] Two months later, he denounced himself to the cantonal police of Zurich because in 1985 he had yet again failed to enrol himself in a number of civil defence exercises and courses.[19] In January 1986, the district attorney of Zurich filed charges against Schnyder, calling for seven days of imprisonment.[20] After that, in February 1986, Schnyder once more demanded of the civil defence office of the canton of Zurich that it not assign him to tasks related to what he called 'war civil defence (*Kriegszivilschutz*)' and emphasised again his willingness to serve in disaster relief.[21] In its written reply, the civil defence office of the canton of Zurich denied the request once again but still 'did not want to give up the hope' that Schnyder 'as a physician would find the way to realise that the categorical rejection of Swiss civil defence was based on knee-jerk thinking (*Kurzschlussdenken*) arising from an all-encompassing fear'.[22]

The prosecution, however, ran its course. On 11 March 1986, the trial against Schnyder at the district court of Zurich took place. Schnyder invited friends, sympathisers and reporters to the trial. He also organised a press conference and a lunch they could attend and sent them documentation about his case when requested.[23] The editor of the newsletter of the West German section of the IPPNW regretted that he was not able to attend the trial in Zurich but wished Schnyder 'big media attention'. Moreover, he planned on using Schnyder's documentation about conscientious objection in Switzerland for an article.[24] This shows that the national sections of the IPPNW could rely on a transnational network in order to exchange ideas and experiences. At the trial, Schnyder was morally supported by well-known Swiss intellectual Max Frisch.[25] Swiss newspapers published various articles about the trial.[26] They provoked controversial reactions, as several letters to the editor illustrate.[27] Schnyder himself pleaded either for a verdict of acquittal or for a prison sentence of 30 days, the latter leading automatically to his exclusion from civil defence service.[28] In sentencing, however, the court imposed a financial penalty on Schnyder, as in an earlier conviction.[29] But this was not the end of Schnyder's attempt to escape nuclear civil defence. Another refusal to serve in 1987 was finally sanctioned with 30 days of imprisonment and, with this, the exclusion criterion was achieved.[30] Schnyder could serve his time in quasi-detention.[31] Finally, in June 1987, the local council of his residential community excluded Schnyder officially from civil defence service.[32]

Schnyder was not the only person who tried to use his trial to get public attention for his political cause. In 1988, for example, 'Crazy is normal' was published, which was the complete documentation of a trial at the cantonal court of Schaffhausen of seven men who refused to serve in the civil defence forces. The publication contained the defence arguments that had been submitted during the trail, including statements by Thomas Schnyder; the social democratic politician and co-founder of the 'Group Switzerland without Army (*Gruppe Schweiz ohne Armee*)' Andreas Gross; the renowned psychoanalyst Paul Parin; and the filmmaker and publicist Alexander J. Seiler, who himself also refused to do civil defence service.[33] Another case that gained widespread attention was that of the journalist Daniel Glass, who himself had also demanded a harsher penalty at a police court in order to be excluded from civil defence service more quickly. While in other cases, the judges had denied these demands or the state attorneys had appealed similar convictions, Glass's case in March 1986 was the first one in which a judge had ruled in favour of such a request. In order to gain publicity for this precedent, Glass sent the opinion of the court to fellow journalists and other conscientious objectors.[34]

Thus, members of PSR/IPPNW Switzerland as well as other conscientious objectors of civil defence used their trials to win the public and the media for their political cause. By promoting and publicising their trials and inviting the press, sympathisers and well-known public figures, they wanted to create sympathy for the idea of conscientious objection in the public sphere. In order to gain media attention, they staged their resistance in court as well as in happenings, showcase events and publications in new ways. These unconventional forms of protest took inspiration from forms of political action developed by new social movements (Albrecht et al. 1988: 12) emerging since the mid-1960s and especially in the wake of '1968' (Schaufelbuehl 2009; Skenderovic and Späti 2012).

Another form of resistance to the Coordinated Medical Services initiated by members of PSR Switzerland was the launching of a public campaign entitled 'Health services call for peace'. Apart from physicians, also nurses, physiotherapists and other health workers supported the call, which was popularised at a press conference in May 1984 in Bern. At that moment in time, 800 people had already signed the call, 500 of them people working in health services.[35] People signing the call did not want to be part of any preparations for nuclear war in health care and did not want to attend any training in war medicine, and they demanded that the courses and exercises in disaster medicine were limited to civil accidents and catastrophes. Furthermore, the signatories of the call stated their solidarity and support for everyone who refused to attend such events. They also refused to accept the newly introduced mandatory registration and conscription of health workers into the Coordinated Medical Services.[36] As already mentioned, this conscription was the authorities' attempt to install a mandatory service for women in particular, who were not obligated to serve under the Swiss federal constitution. The protest call was supported by the Swiss Society for Social Healthcare, the Zurich branch of the Association of Independent Physicians and the Swiss Syndicate of Public Services, a trade union.[37] The protest call reached and mobilised a great number of working medical and nursing staff who themselves were not members of PSR/IPPNW Switzerland. Therefore, it provided a good opportunity for the organisation to reach out to the broader public and gave it a chance to demonstrate how Swiss total national defence directly affected people's everyday lives.

It is noteworthy that, in their argumentation, members of PSR/IPPNW Switzerland always made recourse to their individual morals and ethics. An example of this is the fundamental criticism of the disaster and war medicine approach that was to be employed in the Coordinated Medical Services in the event of a nuclear war.[38] The physician and board member of PSR/IPPNW Switzerland Wolfgang Lauterburg, co-initiator of the abovementioned protest call, prominently criticised the Coordinated Medical Services' inherent militarist logic and its undifferentiated equation of civil emergency with war medicine, and, to that end, the constant conflation of civil and military structures and of the states of peace and war (Lauterburg 1988). He held that the principles of disaster or war medicine were incompatible with his ethics as a physician, as these principles relied on the concept of triage, that is, the sorting of injured people according to the severity of their injuries. Whereas medical ethics would usually require the provision of help to those people who are most injured first, the concept of triage necessitated that so-called hopeless cases be sorted out in order to concentrate resources on those people who had good chances of survival. Thus, triage decisions required a problematic reversal of common medical ethics, as the principles of individual medicine would not apply. Lauterburg and other critics of the Coordinated Medical Services felt that it was against their individual conscience as physicians to work in nuclear civil defence and make such decisions.

The recourse to individual morals and ethics was also at the heart of the new physician's oath that well-known international members of IPPNW signed and that started with the phrase: 'As a matter of individual conscience, [...]'.[39] Likewise, the so-called Frankfurt Declaration (*Frankfurter Erklärung*) of the West German section of the IPPNW, issued in May 1982, referred to conscientious reasons as physicians (Kemper 2014: 327–328). The Frankfurt Declaration functioned as a transnational element, as physicians from Switzerland like Thomas Schnyder also signed it and used it as an argument to explain why they refused to serve.[40] By claiming that they were undergoing a 'crisis of conscience (*Gewissensnotstand*)', members of PSR/IPPNW Switzerland and other conscientious objectors on trial were invoking the conscience of the individual as a legal argument.[41] In these cases, as well as in cases where men refused to serve in the militia army, the crisis of conscience served as grounds for justification beyond the law to explain why someone refused to serve (Schweizer 2017: 24–28).

In sum, in both their plea statements and their public statements, the members of PSR/IPPNW Switzerland often referred to their individual conscience and their ethics as physicians in order to explain why they refused to serve in nuclear civil defence and why they rejected the methods and principles of disaster and war medicine. Their personal commitment and dedication, as well as their willingness to accept severe personal consequences such as imprisonment, gave the physicians' protest a high credibility. By invoking their social authority as physicians, the members of PSR/IPPNW Switzerland made recourse to the traditional subject of the conscientious physician who acts to the best of his knowledge and belief. This form of subjectivisation was an effective discursive strategy, the aim of which was twofold: to gain the normative clout and moral authority needed to bolster resistance within the population on the one hand, and to circumvent being branded as a communist or subversive on the other hand.

Hence, on a discursive level, PSR/IPPNW Switzerland's criticism of the Coordinated Medical Services was based on what the historian Susanne Schregel (2009) with regard to the West German Peace Movement has called 'politics of subjectivity', which enabled the peace activists to express their individual emotions of fear. PSR/IPPNW Switzerland's resistance to the Coordinated Medical Services, however, was less emotionally grounded. Rather, with its 'politics of subjectivity', it propagated a moral stance that claimed that individual moral action mattered and might change politics. To state that the actions of each and every person could make a difference was an extremely empowering vision in an era of the Cold War, which seemed to be dominated by feelings of helplessness and hopelessness (Kemper 2012; Marti 2017). In short, the alternative vision publicly promoted by PSR/IPPNW Switzerland was that a world without nuclear war was possible if people took individual moral decisions to listen to their own conscience.

In contrast to the sociotechnical imaginary of total national defence, which referred to the collective of the 'Swiss people', the collective in PSR/IPPNW Switzerland's counter-vision was the sum of acting individuals with a strong moral compass. And whereas individuals had to bow to the norms and requirements of the Cold War within the sociotechnical imaginary of total national defence, individuals in PSR/IPPNW Switzerland's alternative vision were right to oppose these norms and to take ethical decisions based on individual moral grounds. In the subject of the conscientious physician, this vision found its leading figure.

Civil Defence Critics: Between Public Enemies and Public Support

PSR/IPPNW Switzerland's resistance to the Coordinated Medical Services was effective as publicity, and therefore, was a thorn in the authorities' eye. They responded with hard accusations and legally dubious measures. In the autumn of 1983, the surgeon general of the Swiss armed forces and coordinating leader of the Coordinated Medical Services at the federal level, for example, deliberately dispatched military surgeons to attend a meeting organised by PSR/IPPNW Switzerland and fight their views. In the official minutes, the surgeon general's opinion of PSR/IPPNW Switzerland is recorded as follows: 'This organisation basically calls for an abolition of civil defence, the Coordinated Medical Services, and total national defence. This would be [...] the destruction of the nation'.[42] Thus, members of the armed forces were instructed to infiltrate PSR/IPPNW Switzerland's events because its view was officially regarded as being dangerous and subversive. Likewise, a military surgeon and member of the Federal Commission for Radiation Protection wrote an article 'against defeatism in disaster medicine' in the influential liberal newspaper *Neue Zürcher Zeitung* in September 1982, rejecting the criticism that the maxims of disaster medicine as envisioned in the event of a nuclear war contradicted the Hippocratic oath.[43]

With that said, it is not surprising that PSR/IPPNW Switzerland was under surveillance by Switzerland's state security. In their files, state security declared PSR/IPPNW Switzerland to be a pro-Soviet 'front organisation'.[44] Swiss state security's secret observations of tens of thousands of innocent people and organisations only came to light in November 1989 and resulted in one of the biggest scandals in Switzerland's recent history (Liehr 2014). But in the first half of the 1980s, Swiss authorities still stubbornly kept to their Cold War enemy stereotypes and total national defence feasibility fantasies. Anyone who like the members of PSR/IPPNW Switzerland publicly questioned Swiss total national defence organisations was basically seen and treated as an enemy of the state. Thus, PSR/IPPNW Switzerland is an example of how, in Cold War Switzerland, state security targeted groups and organisations from civil society that operated democratically but were opposed to the total national defence imaginary.

Besides the authorities' suspicion and surveillance, PSR/IPPNW Switzerland's protest actions against nuclear civil defence and the Coordinated Medical Services, respectively, also provoked reactions from the public. Apart from the already mentioned letters to the editor, people occasionally wrote personal letters to public figures who represented PSR/IPPNW Switzerland. Some of these correspondents uttered mixed or critical feelings about PSR/IPPNW Switzerland's resistance. A physician and colleague of Thomas Schnyder, for example, responded to an article Schnyder had published in the journal *Social Medicine* with regard to his decision to refuse to serve in the Coordinated Medical Services.[45] This correspondent held that the 'head-in-the-sand politics' of physicians was not productive. In his opinion, it was better to have civil defence than no prevention measures at all. For him, civil defence did not downplay the danger of war. Rather, confrontation with it, for instance, through the civil defence leaflet in the phone book, might make people think about how such a war could be prevented.[46]

In most of these letters, however, people expressed their support and described how PSR/IPPNW Switzerland's actions had strengthened their own political commitment to stand up and fight for peace. Several men wrote that, after they had read articles about PSR/IPPNW Switzerland's protest actions, they now felt ready to act themselves and would refuse to serve or refuse to attend special training in civil defence.[47] Another man thanked them for their help, as he planned to refuse to accept a managerial post in the civil defence service.[48] One man wrote that he would like to act in a similar way but lacked the courage because he worked as an official for a cantonal authority.[49] Yet another man wrote that he now felt encouraged to publish a leaflet criticising a nearby planned civil defence building.[50] Not only men but also women wrote to tell how similar their own mindset was, and congratulated the courage and the responsible, consistent and brave attitude they demonstrated.[51] Even people from abroad wrote letters. A man from the West German city of Nuremberg, for example, expressed his sympathy and support for Thomas Schnyder's decision not to serve.[52] This once again illustrates that civil defence criticism in the last decade of the Cold War was a transnational phenomenon, and that there was an active exchange of information and ideas.

To be sure, these letters are far from being a representative sample of what Swiss men and women thought about civil defence in the 1980s, and it is likely that many letters came from people that stood close to the peace movement and thus shared the rather negative attitude towards Switzerland's total national defence system. Still, they are rare and insightful sources for studying how the total national defence imaginary and alternative visions articulated by civil defence critics like PSR/IPPNW Switzerland resonated in Cold War Switzerland of the 1980s.

In general, in the course of the 1980s, criticism against the armed and the civil defence forces enjoyed more and more popularity. At the end of the decade, a referendum demanding the abolishment of the Swiss army launched by the 'Group Switzerland without Army' unexpectedly received a third of all the votes, clearly showing that identification with the key symbol of Switzerland's ability to defend its home ground was eroding (Degen 2017). Likewise, in a population survey in 1989, less than twenty per cent of the respondents thought that Swiss civil defence would be able to safeguard the population in the event of a nuclear war.[53] But the authorities' optimistic belief in the feasibility of survival underlying Switzerland's total national defence also began to crumble slightly after a scientific study commissioned by the Central Office for Defence had clearly shown how extremely disastrous a nuclear war really would be (Berger Ziauddin and Marti 2020).

In sum, the reactions to PSR/IPPNW Switzerland's resistance against nuclear civil defence illustrate how divided Swiss society was at the end of the Cold War. On the one hand, there were still many hard-nosed 'Cold Warriors' in powerful positions and institutions, and, on the other hand, there were younger people, in particular, who no longer believed in the imaginary of Switzerland's total national defence. While Swiss authorities still tried to denounce PSR/IPPNW Switzerland's counter-vision as defeatism, this accusation could not muster the same authority as in earlier decades of the Cold War. In fact, the movement of conscientious objection in the 1970s and the 1980s which PSR/IPPNW Switzerland was part of paved the way for the introduction of an alternative service based on conscientious reasons shortly after the end of Cold War (Koller 2008). Thus, in Cold War Switzerland of the 1980s, there was no homogenous sentiment towards PSR/IPPNW Switzerland's criticism of nuclear civil defence (Kemper 2016: 280). But this does not mean that the criticism of the sociotechnical imaginary of total national defence made by PSR/IPPNW Switzerland and other protest groups did not resonate with the Swiss public. Rather, this resistance was fiercely contested within Swiss society and, therefore, gained widespread public and media attention.

Conclusion

In Switzerland, Cold War perceptions and dichotomies developed a remarkable normative power. The threat of a nuclear war between the East and the West resulted in the building of a strong total national defence. The Coordinated Medical Services, which had been set up in the mid-1960s and were operational at the beginning of the 1980s, were an emblematic total national defence organisation created as one of Switzerland's prevention measures against a nuclear war. It operated with a hybrid organisational structure involving military and civil tasks and structures that blurred the boundaries between the states of war and peace. For the Coordinated Medical Services to function, they relied on the mobilisation of institutions of federal levels and of the nation as a whole, with private organisations playing a vital role. Women, who did not have the obligation to serve under the Swiss federal constitution, were urged to volunteer or forced to serve by newly introduced cantonal laws. Within the Coordinated Medical Services, they were expected to take on their traditional gender roles in nursing and caregiving.

Through organisations like the Coordinated Medical Services, Swiss authorities propagated an imaginary that, with a massive collective effort, Switzerland could survive a nuclear war. But during the 1980s, civil society protest and criticism of total national defence gained strength, leading to the erosion of core components of this imaginary towards the end of the decade. This resistance was at once part of transnational political actions for nuclear disarmament and peace as well as part of Cold War Switzerland's domestic conflicts about the armed and civil defence forces and their dominant role in Swiss society. The Coordinated Medical Services, too, had to face resistance, the core of which was PSR/IPPNW Switzerland. The members of PSR/IPPNW Switzerland refused to be part of all measures and institutions concerned with nuclear civil defence, and they effectively fought against it. Their trials as a result of their conscientious objection, as well as their political campaigns, were staged as a public protest theatre.

In the authorities' perception, PSR/IPPNW Switzerland was fostering collective resistance, as its members refused to bow to the sociotechnical imaginary of Swiss total national defence. State security surveillance shows that PSR/IPPNW Switzerland was regarded as a subversive organisation opposed to democracy and the rule of law. Therefore, PSR/IPPNW Switzerland's resistance to the Coordinated Medical Services and nuclear civil defence, respectively, sheds light on how strong the role played by Cold War norms and requisites of security still was in Switzerland of the 1980s, and it demonstrates how difficult it was to build up democratic opposition against the prevailing imaginary of total national defence.

Under these circumstances, PSR/IPPNW Switzerland had to find a way for its resistance to gain legitimacy among the Swiss public. Subjectivisation was an effective discursive strategy that its members used to justify their protest and criticism. Centred around the figure of the conscientious physician, the members of PSR/IPPNW Switzerland promoted political actions as individual moral decisions. By openly propagating these subjectivised forms of resistance, they wanted to encourage more people do what they did, that is to follow their own conscience and to resist nuclear civil defence. The personal letters that several publicly known members received illustrate that some people actually felt more encouraged to take on resistive behaviour themselves after they had learned about PSR/IPPNW Switzerland's actions. In this way, the strategy of subjectivisation enabled PSR/IPPNW Switzerland to depict its criticism and protest not as subversive resistance but as the result of conscientiously acting individuals taking similar moral decisions to stand up for peace.

Contrary to what Swiss authorities and Swiss state security insinuated, the members of PSR/IPPNW Switzerland did not fundamentally oppose the state or state institutions. Rather, their subjectivised forms of resistance were a strategic reaction to Switzerland's imaginary of total national defence: whereas this imaginary urged the 'Swiss people' to adhere to imposed Cold War roles and norms, PSR/IPPNW Switzerland called on the Swiss population to act as conscientious subjects who had the right to make individual ethical decisions. Thus, PSR/IPPNW Switzerland's alternative vision that a world without nuclear war was possible relied more on the moral strength of human beings than on the instigated pressures of Cold War politics.

Notes

1. Archive FOPH, 18.2.60, Totaler Sanitätsdienst und umfassende AC-Schutzmassnahmen, Antrag des EMD, 8 February 1968. All translations of quotes in this chapter are by the author.
2. CH-BAR#E5540E#1984/63#1*, Koordinierter Sanitätsdienst KSD Konzept, 1 December 1980.
3. CH-BAR#E5540E#1984/63#1*, Koordinierter Sanitätsdienst KSD Konzept, 1 December 1980.
4. CH-BAR#E5540E#1994/14#64*, Protokoll der Sitzung des Büros Ausschuss San D, 28 November 1973.
5. CH-BAR#E5540E#1984/63#3*, Kanton 'Rigi': KSD-Dokumentation (Provisorische Ausgabe), 1 January 1981.

6. See the manuscripts and articles in CH-BAR#E4113A#1982/54#38*.
7. CH-BAR#E4113A#2000/390#73*, Standard-Referat mit Folien, 1 May 1979, Standard-Referat mit Folien, 1 October 1980, and Standardreferat Koordinierter Sanitätsdienst Konzept mit Folien, 25 July 1984.
8. CH-BAR#E5540E#1994/14#291*, Koordinierter Sanitätsdienst: Informationskonzept, 17 June 1974, emphasis in original.
9. CH-BAR#E5540E#1994/14#59*, letter from Oberstdivisionär Käser to W. König, 16 February 1973, Katastrophenmedizin WS 1972/73 Zürich, without date, and Katastrophenmedizin, Katastrophenorganisation im Kantonsspital und im Raume Basel, 14 April 1973.
10. CH-BAR#E5540F#1989/97#16*, Protokoll der Sitzung des Büros des Ausschusses Sanitätsdienst, 21 September 1978.
11. CH-BAR#E5540F#1989/97#27*, Referat vor Mitgliedern der Sektionen Schaffhausen des Schweizerischen Zivilschutz-Verbandes und der Schweizerischen Offiziersgesellschaft, 29 December 1978, emphasis in original.
12. CH-BAR#E5540F#1989/97#16*, Protokoll der Sitzung des Büros des Ausschusses Sanitätsdienst, 21 June 1978.
13. SocArch, Ar 526.40.6, Problem der Zivilschutzverweigerer wird vorläufig verdrängt, in *Basler Zeitung*, 16 February 1985.
14. SocArch, Ar 526.40.6, Chronologie der Laufbahn im Zivilschutz von T. Schnyder, 3 February 1986.
15. SocArch, Ar 526.40.6, letter from T. Schnyder to Amt für Zivilschutz des Kantons Zürich, 26 May 1984.
16. SocArch, Ar 526.40.6, letter from Amt für Zivilschutz des Kantons Zürich to T. Schnyder, 6 July 1984.
17. SocArch, Ar 526.40.6, letter from T. Schnyder to P. Widemann, 17 April 1985.
18. SocArch, Ar 526.40.6, Chronologie der Laufbahn im Zivilschutz von T. Schnyder, 3 February 1986.
19. SocArch, Ar 526.40.6, letter from T. Schnyder to P. Bliggenstorfer, 1 June 1985.
20. SocArch, Ar 526.40.6, Anklageschrift der Bezirksanwaltschaft Zürich, 29 January 1986.
21. SocArch, Ar 526.40.6, letter from T. Schnyder to B. Hersche, 3 February 1986.

22. SocArch, Ar 526.40.6, letter from B. Hersche to T. Schnyder, 7 February 1986.
23. SocArch, Ar 526.40.6, letter from T. Schnyder to friends and sympathisers, 26 February 1986, Einladung zu Pressegespräch, 26 February 1986, and Antworttalon Prozess Kriegszivilschutzverweigerung, without date.
24. SocArch, Ar 526.40.6, letter from M. Roelen to T. Schnyder, 3 March 1986.
25. SocArch, Ar 526.40.6, Ist Zivilschutz (noch) sinnvoll? in *AZ*, 12 March 1986.
26. See, e.g., SocArch, Ar 526.40.6, Bei Atomalarm Kragen hochstellen! in *Weltwoche* No. 14, 3 April 1986, and Ist Zivilschutz (noch) sinnvoll? in *AZ*, 12 March 1986.
27. SocArch, Ar 526.40.6, Leserbriefe betreffend 'Protestaktion', in *Weltwoche* No. 16, 17 April 1986.
28. SocArch, Ar 526.40.6, Plaedoyer zur Verweigerung des Kriegszivilschutzes von Dr. med. Thomas Schnyder, without date.
29. SocArch, Ar 526.40.6, Urteil des Bezirksgerichts Zürich, 12 March 1986.
30. SocArch, Ar 526.40.6, Strafbefehl der Bezirksanwaltschaft Meilen, 3 February 1987.
31. SocArch, Ar 526.40.6, Strafvollzug in Form der Halbgefangenschaft, 15 May 1986.
32. SocArch, Ar 526.40, Auszug aus dem Protokoll des Gemeinderates Zollikon, 3 June 1987.
33. SocArch, Ar 526.40.6, Werner Brandenberger: Plädoyer für das Gewissen, without date, and 'Das Verrückte ist normal'. Ueberlegungen zur beiliegenden Konzeptmaquette einer Publikation zum Schaffhauser Zivildienstverweigerer-Prozess 1988, 15 October 1988.
34. SocArch, Ar 526.40.6, letter from D. Glass to the press and the radio and people interested in conscientious objection, 15 March 1986, and Zivilschutz-Verweigerer wollte Haft statt Busse: 30 Tage unbedingt, in *Basler Zeitung* No. 65, 18 March 1986.
35. SocArch, Ar 526.40.6, Aufruf aus dem Gesundheitswesen für Frieden: Einladung zu einer Pressekonferenz, 12 May 1984.
36. SocArch, Ar 526.40.6, Aufruf aus dem Gesundheitswesen für Frieden: Erklärung, without date.

37. SocArch, Ar 526.40.6, Aufruf 'Gesundheitswesen für Frieden', in *Der Bund*, 29 May 1984.
38. See, e.g., CH-BAR#E5540E#1984/63#5*, Koordinierter Sanitätsdienst (KSD), Behelf Basisspital (BBS), Ausgabe 1980, May 1980.
39. SocArch, Ar 526.40.6, A New Physician's Oath, without date.
40. SocArch, Ar 526.40.6, Warum ich als Arzt den Dienst im Zivilschutz verweigere, May 1984.
41. SocArch, Ar 526.40.6, Zivilschutzverweigerungsprozess gegen Dr. med. Hansueli Leuzinger am Obergericht des Kantons Zürich, 31 August 1989, and Werner Brandenberger: Plädoyer für das Gewissen, without date.
42. CH-BAR#E5540F#1989/97#11*, Protokoll der Ausschuss-Sitzung, 22 November 1983.
43. CH-BAR#E4390C#1997/14#525*, Medizinische Vorsorge gegen den Atomkrieg, in *Neue Zürcher Zeitung* No. 207, 7 September 1982.
44. SocArch, Ar 526.20.12, Frontorganisationen: International Physicians for the Prevention of Nuclear War IPPNW, 6 July 1989.
45. SocArch, Ar 526.40.6, Warum ich den Dienst beim Zivilschutz als Arzt verweigere, in *Soziale Medizin* 11/7, October 1984.
46. SocArch, Ar 526.60.2, letter from M. Ryffel to T. Schnyder, 11 November 1984.
47. SocArch, Ar 526.40.6, letter from A. Fritschi to T. Schnyder, without date, and letter from P. Gmür to T. Schnyder, without date.
48. SocArch, Ar 526.40.6, letter from M. Righetti to T. Schnyder, 18 December 1986.
49. SocArch, Ar 526.40.6, letter from T. Schnyder, 13 March 1986.
50. SocArch, Ar 526.40.6, letter from B. Lustenberger to T. Schnyder, 10 April 1986.
51. SocArch, Ar 526.40.6, letter from A. Kaspar to T. Schnyder, 21 April 1986, and letter from M. Vögeli to T. Schnyder, 27 April 1986.
52. SocArch, Ar 526.40.6, letter from H. Schöckmann to T. Schnyder, 13 April 1987.
53. Zivilschutz ist nie ein absoluter Schutz, in *Tages-Anzeiger*, 23 June 1990.

References

Unpublished Sources

Archives of the Federal Office of Public Health, Bern Liebefeld (Archive FOPH)

- Archive FOPH, 18.2.60, Totaler Sanitätsdienst und umfassende AC-Schutzmassnahmen.

Swiss Federal Archives, Bern (CH-BAR)

Holdings E 4113A*: Dienst für kriegsnotrechtliche Sonderfragen: Zentrale Ablage (1963–1983) (1914–1997)

- CH-BAR#E4113A#1982/54#38*, Az. 281.1, ASAN, 1963–1972.
- CH-BAR#E4113A#2000/390#73*, Az. 281.1, Ausschuss Sanitätsdienst, 1976–1984.

Holdings E 4390C*: Bundesamt für Zivilschutz: Zentrale Ablage (1976–2002) (1928–1999)

- CH-BAR#E4390C#1997/14#525*, Az. 137.110, Allgemeines, 1982.

Holdings E 5540E*: Bundesamt für Sanität: Zentrale Ablage (1979–1982) (1922–1988)

- CH-BAR#E5540E#1984/63#1*, Az. 1.1280.1, Konzept vom 1.12.1980, 1980.
- CH-BAR#E5540E#1984/63#3*, Az. 1.1280.12, KSD-Dokumentation Fachkurse 'Sanitätsdienst III', ständige Weisungen, Kanton 'RIGI', Prov. Ausgabe Stand der Erkenntnisse 1.1.1981, 1981.
- CH-BAR#E5540E#1984/63#5*, Az. 1.1280.4, Behelf Basisspital (BBS) Ausg. 1980, 1980.
- CH-BAR#E5540E#1994/14#59*, Az. 1.1200, Katastrophen(-Medizin), 1973.
- CH-BAR#E5540E#1994/14#64*, Az. 1.1280, Ausschuss Sanitätsdienst Stab für Gesamtverteidigung, 1973.
- CH-BAR#E5540E#1994/14#291*, Az. 1.1280, Ausschuss Sanitätsdienst Stab für Gesamtverteidigung, 1974.

Holdings E 5540F*: Bundesamt für Sanität: Zentrale Ablage (1983–1995) (1928–1997)

- CH-BAR#E5540F#1989/97#11*, Az. 12.2, Ausschuss-Sitzungen 1973–77, 1979–1984.
- CH-BAR#E5540F#1989/97#16*, Az. 12.2, Büro KSD, 1973–1982.
- CH-BAR#E5540F#1989/97#27*, Az. 12.2, KSD September 1978–März 1979 (Protokolle), 1978–1979.

Swiss Social Archives, Zurich (SocArch)

Holdings Ar 526, ÄrztInnen für soziale Verantwortung und zur Verhütung eines Atomkrieges (PSR/IPPNW-Schweiz)

- SocArch, Ar 526.20.12, Diverse Akten I, 1985–2009.
- SocArch, Ar 526.40.6, Dossier Zivilschutz-Verweigerung, 1983–1990.

– SocArch, Ar 526.60.2, Korrespondenz Th. Schnyder, 1983–1995.

Literature

Albrecht, Peter et al. (Autorenkollektiv). 1988. Einleitung. In *Schutzraum Schweiz: Mit dem Zivilschutz zur Notstandsgesellschaft*, eds. Peter Albrecht et al. (Autorenkollektiv), 9–15. Bern: Zytglogge.

Becker-Schaum, Christoph et al., eds. 2016. *The nuclear crisis: The arms race, Cold War anxiety, and the German peace movement of the 1980s*. New York and Oxford: Berghahn.

Bein, Thomas, and Rudolf Epple. 1986. Die Friedensbewegung heute: Rahmenbedingungen und Tendenzen. In *Handbuch Frieden Schweiz*, ed. Forum für praxisbezogene Friedensforschung, 9–122. Basel: Genossenschaft Z-Verlag.

Berger Ziauddin, Silvia. 2017. Superpower underground: Switzerland's rise to global bunker expertise in the Atomic Age. *Technology and Culture* 58: 921–954.

Berger Ziauddin, Silvia and Sibylle Marti. 2020. Life after the bomb: Nuclear fear, science, and security politics in Switzerland in the 1980s. *Cold War History* 20: 95–113.

Bondolfi, Sibilla. 2012. *Wehrpflicht und Geschlechterdiskriminierung: Verfassungsrechtliche und völkerrechtliche Anforderungen an die Wehrpflicht im Vergleich zum Modell Israel*. Zürich: Schulthess.

Brändli, Sebastian. 2012. Gesundheitswesen. In *Historisches Lexikon der Schweiz (HLS)*, version from 6 December 2012, https://hls-dhs-dss.ch/de/articles/016593/2012-12-06/ [accessed 22 November 2019].

Bundesgerichtsentscheid. 1989. *BGE 115 Ia 277* – Katastrophenfall, 3 May 1989.

Conze, Eckart et al., eds. 2017. *Nuclear threats, nuclear fear: Cold War of the 1980s*. New York: Cambridge University Press.

Degen, Bernard. 2009. Die totale Verteidigungsgesellschaft. In *Krieg*, eds. Christoph Maeder et al., 89–105. Zürich: Seismo.

Degen, Bernard. 2017. Gruppe Schweiz ohne Armee (GSoA). In *Historisches Lexikon der Schweiz (HLS)*, version from 24 August 2017, https://hls-dhs-dss.ch/de/articles/015312/2017-08-24/ [accessed 22 November 2019].

Epple, Ruedi. 1986. Die schweizerische Friedensbewegung: Alte Wurzeln – Neue Blüten. *Jahrbuch der Schweizerischen Vereinigung für politische Wissenschaft* 26: 193–210.

Kemper, Claudia. 2012. Psychologische Abrüstung: Psychotherapeuten in der westdeutschen Friedensbewegung der frühen 1980er Jahre. In *Das Selbst zwischen Anpassung und Befreiung. Psychowissen und Politik im 20. Jahrhundert*, eds. Maik Tändler and Uffa Jensen, 168–185. Göttingen: Wallstein.

Kemper, Claudia. 2014. 'Wir werden euch nicht helfen können': Medizinische Ethik und atomare Bedrohung zu Beginn der 1980er Jahre. In *Den Kalten Krieg denken. Beiträge zur sozialen Ideengeschichte seit 1945*, eds. Patrick Bernhard and Holger Nehring, 307–335. Essen: Klartext.

Kemper, Claudia. 2016. *Medizin gegen den Kalten Krieg: Ärzte in der anti-atomaren Friedensbewegung der 1980er Jahre*. Göttingen: Wallstein.

Koller, Christian. 2008. Der lange Weg zum 'zivilen Ersatzdienst' in der Schweiz. In *Ich dien' nicht! Wehrdienstverweigerung in der Geschichte*, eds. Christian Th. Müller and Dierk Walter, 227–242. Berlin: dissertation.de – Verlag im Internet.

Lauterburg, Wolfgang. 1988. Aufrüstung mit anderen Mitteln: Der Koordinierte Sanitätsdienst und die Katastrophenmedizin. In *Schutzraum Schweiz: Mit dem Zivilschutz zur Notstandsgesellschaft*, eds. Peter Albrecht et al. (Autorenkollektiv), 199–217. Bern: Zytglogge.

Liehr, Dorothee. 2014. *Skandal und Nation: Politische Deutungskämpfe in der Schweiz 1988–1991*. Marburg: Tectum Verlag.

Marti, Sibylle. 2015. Den modernen Krieg simulieren: Imaginationen und Praxis totaler Landesverteidigung in der Schweiz. In *Das Imaginäre des Kalten Krieges: Beiträge zu einer Kulturgeschichte des Ost-West-Konfliktes in Europa*, eds. David Eugster and Sibylle Marti, 243–268. Essen: Klartext.

Marti, Sibylle. 2017. Nuklearer Winter – emotionale Kälte: Rüstungswettlauf, Psychologie und Kalter Krieg in den Achtzigerjahren. In *Der kalte Krieg: Kältegrade eines globalen Konflikts*, eds. Silvia Berger Ziauddin et al., 157–174. Berlin and Zürich: Diaphanes.

Marti, Sibylle. 2020. *Strahlen im Kalten Krieg: Nuklearer Alltag und atomarer Notfall in der Schweiz*. Paderborn: Ferdinand Schöningh.

Meier, Yves and Martin Meier. 2010. Zivilschutz – Ein Grundpfeiler der Schweizer Landesverteidigung im Diskurs. *Schweizerische Zeitschrift für Geschichte* 60: 212–236.

Möcklin, Emanuel. 1998. *Militärdienstverweigerung: Über die Ursachen der Militärdienstverweigerung in der Schweiz*. Lizenziatsarbeit Universität Zürich.

Schaufelbuehl, Janick Marina, ed. 2009. *1968–1978: Ein bewegtes Jahrzehnt in der Schweiz*. Zürich: Chronos.

Schindler, Roxane D. 1997. *Die allgemeine Dienstpflicht*. Zürich: Schulthess.

Schregel, Susanne. 2009. Konjunktur der Angst: 'Politik der Subjektivität' und 'neue Friedensbewegung'. In *Angst im Kalten Krieg*, eds. Bernd Greiner et al., 495–520. Hamburg: Hamburger Edition.

Schweizer, Andrea. 2017. *Die Beratungsstelle für Militärverweigerer: Militärkritik und Obstruktion in der Schweiz 1981–1989*. Master's dissertation, Universität Zürich.

Senn, Hans. 1983. *Friede in Unabhängigkeit: Von der Totalen Landesverteidigung zur Sicherheitspolitik*. Frauenfeld: Huber & Co. AG, Presseverlag.

Skenderovic, Damir and Christina Späti. 2012. *Die 1968er-Jahre in der Schweiz: Aufbruch in Politik und Kultur*. Baden: Hier und Jetzt.

Tanner, Jakob. 1999. Switzerland and the Cold War: A neutral country between the 'American Way of Life' and 'Geistige Landesverteidigung'. In *Switzerland and War*, eds. Joy Charnley and Malcolm Pender, 113–128. Bern: Peter Lang.

Winet, Ruedi, ed. 1991. *Militär- und Zivilschutzverweigerung in der Schweiz: Ein Handbuch*. Zürich: Limmat.

Wittner, Lawrence S. 2003. *Toward nuclear abolition: A history of the world nuclear disarmament movement, 1971 to the present*. Stanford: Stanford University Press.

Ziemann, Benjamin. 2009. A quantum of solace? European peace movements during the Cold War and their elective affinities. *Archiv für Sozialgeschichte* 49: 351–389.

CHAPTER 9

Remembering Desirable Futures? Civil Defence Memories and Everyday Life in Sweden and the UK

Marie Cronqvist and Matthew Grant

How do you write the history of people trying to remember how they imagined desirable or undesirable futures? Analysing civil defence as an imaginary involves conceptualising civil defence as a technology, but it also reminds us of the fact that civil defence was above all a way of mapping out a future for states, societies and individuals. As a future-oriented activity—mapping out and rehearsing the future in the present—it was at the same time abstract and firmly rooted in space and time. When people are looking back at their own experiences of Cold War civil defence from the vantage point of the 2000s, a layer of complexity is added because the act of remembering the future is removed, at least in part, from its anchorage in material experience. We now know that the

M. Cronqvist (✉)
Department of Communication and Media, Lund University, Lund, Sweden
e-mail: marie.cronqvist@kom.lu.se

M. Grant
Department of History, University of Essex, Colchester, UK

M. Cronqvist et al. (eds.), *Cold War Civil Defence in Western Europe*,
https://doi.org/10.1007/978-3-030-84281-9_9

undesirable future of nuclear war never happened—at least not within the time frame that in the history books is called 'the Cold War'.

The aim of this chapter is to explore how the concept of sociotechnical imaginaries (Jasanoff and Kim 2015) can be used to examine civil defence as remembered. The main focus will not be the memory work of civil defence museums or monuments, but instead oral histories testifying to the entanglement of civil defence in everyday life. Is civil defence viewed as having continued validity, or as essentially defunct as a technology? Can we identify major shifts between how civil defence was experienced during the Cold War and how it is remembered afterwards, and can we use the concepts of extension, embedding or resistance to unpack these possible shifts?

The chapter employs a historical ethnography approach, using primarily oral history sources such as interviews and questionnaires collected between 2006 and 2012 with examples from Sweden and the UK. We will use three lenses or themes in order to discuss the memory work of civil defence: *localities*, *temporalities* and *mediations*. Our approach, contrasted with much scholarship on sociotechnical imaginaries, thus attempts a more 'bottom-up' perspective. We aim to provide a refined understanding of *how* civil defence enters memory as a 'collectively held, institutionally stabilized, and publicly performed' vision of a desirable future 'animated by shared understandings of forms of social life and social order attainable through, and supportive of, advances in science and technology' (Jasanoff and Kim 2015: 4).

Sociotechnical Imaginaries and Civil Defence Memories

We believe the sociotechnical imaginaries framework could also be employed when exploring civil defence and everyday life, but any attempt to do this needs to address two potential problems. First, it seems to conceptualise cultural and social change teleologically: technologies are first imagined and then become material. It is a framework, after all, which seeks to explain how technology is understood and circulated by societies and cultures. Secondly, there seems to be a reliance on top-down readings of how technologies are created, circulated and understood: technologies are imagined and circulated by experts and taken up by others within 'culture'. These two potential problems may, however, best be described as a tendency in scholarly works that have so far engaged with the framework. As an overarching theory, we argue that the framework *does* leave space for deeper, less clearly defined ways of understanding the cultural and social uptake of the imaginaries discussed. In particular, the phases of *embedding*, *extension* and *resistance* (see Introduction, this volume) can all be seen as contestable spaces in which meanings are made and remade from below.

A focus on memory allows us to explore how sociotechnical imaginaries can be usefully deployed to interpret everyday understandings of civil defence in ways which nuance the two problems discussed above. As Sheila Jasanoff herself puts it, 'memory work' allows 'novel technoscientific constructs to be most readily naturalised when they fall into line with the way things are remembered as being' (Jasanoff and Kim 2015: 339). At a basic level, people's memories always tend to diverge from any attempt to create neat paths or narratives within the civil defence sociotechnical imaginary. Everyday understandings of civil defence were confused, partial and often idiosyncratic—and so are the memories. This reflects the central, inescapably imaginary nature of civil defence. Civil defence as a *technology* relied on a sense that an imaginable future could be tackled, (hopefully) mastered or (at worst) ameliorated. The nature of the disaster to be tackled, the assumptions about destruction, about human behaviour in the post-attack period, all had to be *imagined* and were so using historical assumptions, culturally informed ideas and insights from the social and psychological sciences.

In rethinking and challenging the problematic teleologies of civil defence sociotechnical imaginaries, we must stress that imaginary scenarios were inflicted by, and mapped onto, the *material* world—the physical geography and built environment—and the materiality of the concrete civil defence technologies that were often only partly visible (physical preparations, bunkers and so on). Jasanoff argues that the proposed framework helps resolve the binary of 'ideas' and the 'material', but examining civil defence memory highlights that this is not a binary at all, that within memory work the *imaginary* and *material* are always linked, and are indeed mutually constitutive to the process of making meaning. It is particularly important to nuance the framework in this way because it allows us to understand the circular, symbiotic nature of the imaginary and the material. Jasanoff herself makes it clear that the very idea of sociotechnical imaginaries was to examine how ideas became a reality within 'cultures, institutions, and materialities'. At times, however, the process can appear rather lineal (Jasanoff and Kim 2015: 322–323). The *imaginary* nature of civil defence, and the nature of memory work, means that any linear understanding of the process by which the imaginary is translated into 'cultures, institutions, and materialities' is highly problematic. That work of imagination was continual, embedded in memory work and formed alongside interactions with and understandings of the material. It was 'always already at work in the work', as Jacques Derrida reminds us in a much-cited quote (Derrida 1986: 71).

One way to think about civil defence and to problematise processes laid out in the sociotechnical imaginaries framework is thus to keep in mind the essential 'fuzziness' of everyday life. Real-life situations are always ambiguous, and people make meaning from their different situations in a variety of different ways depending on identity, social position and cultural experience. In this chapter, we have therefore chosen to highlight this complexity by comparing different national, regional and local contexts. The differences arising out of the empirical material are differences both between and within Sweden and the UK. Memory work has the tendency to sort out and (re)narrativise fuzziness, and the process of recounting memory serves to make order out of chaos, creating new (or alternative) frames of interpretation and creating whole new areas of ambiguity.

Historians have long argued that the circulation of dominant cultural discourses within popular culture in the years between the events remembered and the act of remembering shapes individual memory (Abrams 2010; Assman 2011; Confino 1997). Individual memory, therefore, cannot be entirely separated from the narratives circulating within popular culture either at the time of the events experienced or in the time separating those events and the interview encounter in which they are recalled. Scholars working on 'popular memory' have long argued that individual memory is indelibly shaped by intervening cultural narratives (Dawson 1994), arguing that for memory to be formed, retained and recalled, the 'conditions for its expression' need to exist (Passerini 2003: 238). 'Popular memory' can be defined as an attempt to understand how cultural discourses shape the content and form of what is recalled by individuals, the meanings individuals attach to their own memories and the emotions these memories provoke in them (Dawson 1994). As Penny Summerfield puts it, individuals make sense of their experiences by drawing on:

> generalized, public versions of the aspects of the lives that they are talking about to construct their own particular, personal accounts. This process of life-story telling is crucial to the construction of the subject – in reproducing the self as a social identity, we necessarily draw upon public renderings. (Summerfield 2004: 68)

If no relevant cultural framework is readily available, individual memory breaks down into narrative incoherence or a lack of remembering. Summerfield's popular memory approach is particularly important for understanding the interrelationship between cultural narratives and individual life circumstances (Summerfield 1998).

Our work maintains the position that people remain agents and are not empty vessels awaiting cultural discourse to pour meaning into their minds. Our comparative analysis illustrates different levels of 'fuzziness' of civil defence memory, and of course the intensely *political* nature of civil defence merely added to this complexity. Our case studies highlight that civil defence memory work perhaps reveals more about acts of defiance or forgetting, conflicts and contradictions than it does about any 'collectively held, institutionally stabilized, and publicly performed' vision (Jasanoff and Kim 2015: 4). Perhaps one way of thinking about the fuzziness of everyday civil defence, and the difficulty in reconstructing the meaning civil defence had during the Cold War, is to examine the emotional dimensions of the topic. Civil defence could be seen as a technology aimed at emotional regulation, an attempt to contain and manage *fear* and to encourage a more intellectual acceptance of *security*. The use of civil defence as a regulatory, or governmental, technology aligns with what William Reddy discussed as the 'emotional regime', the prevailing norm about how to 'feel'. Reddy defined the emotional regime as '[t]he set of normative emotions and the official rituals, practices, and 'emotives' that express and inculcate them; a necessary underpinning of any stable political regime' (Reddy 2001: 129). As such, the concept of emotional regime reminds us of a sociotechnical imaginary, meaning a largely top-down structure. It also reminds us of the approach in some previous influential studies on civil defence, such as Guy Oakes' study on what he calls the 'emotional management' of American civil defence drills (Oakes 2004).

More fruitful for the perspective in this chapter, then, is perhaps Reddy's idea of 'emotional navigation', signifying the process whereby people navigate and negotiate their everyday lives (Reddy 2001). The multiple experiences of civil defence can be seen as processes not only of collective emotional navigation through the attempt made by national civil defence cultures to impose an 'emotional regime' on people but also of individual emotional navigation through the complex emotions that confronting civil defence would have thrown up. Concepts such as Reddy's 'emotional navigation' might then be more in line with the sort of bottom-up approach that is needed to fully round out the sociotechnical imaginaries framework. Moreover, emotion, like memory and civil defence, was not a purely individual matter, and within groups involved in civil defence, Barbara H. Rosenwein's idea of 'emotional communities' is equally useful for this reason, as it reminds us not only of the social role of civil defence but also of the possible co-existence of various emotional communities encompassing the whole scale of reactions to civil defence, from enthusiasm, acceptance and obedience to criticism and active resistance (Rosenwein 2006).

Civil Defence Memories in Sweden and the UK

In what follows, we will try to anchor this theoretical discussion in civil defence memories from two different national contexts, Sweden and the UK. The Swedish material consists of written responses to a questionnaire on memories of the Cold War, collected in 2006 by the Folklife Archives with the Scania Music Collections at Lund University in collaboration with Marie Cronqvist. The questionnaire generated approximately 80 responses evenly distributed over generation, gender, social class and geography (though mostly from southern Sweden). Since everyday civil defence culture was at the core of the underlying research project, the questionnaire had a special section devoted to memories of civil defence.[1]

The British material emanates from an oral history project conducted by Matthew Grant and Lindsey Dodd between 2010 and 2012 with people who had served in the Civil Defence Services in various capacities during the 1950s and 1960s (Grant 2019). In-depth interviews with eleven participants, either on their own or in small groups, were conducted. There were seven men and four women who had been mainly younger recruits who served in the late 1950s and 1960s. As members of the Civil Defence Corps, the Auxiliary Fire Service and administrative staff employed by a local council, they provided a mixture of experiences from within civil defence. The semi-structured interview schedule asked them to reflect on their reasons for joining civil defence, their experiences within the organisation, their thoughts about the role of civil defence in a war and why they left the service. A central part of the interview was about the ability of civil defence to save lives in a nuclear war. Therefore, we get a sense of how they imagined civil defence as a *technology* even though they did not use the term.

Using these two studies, we examine the memories of civil defence in order to better address the relationship between the imaginary nature of civil defence and the 'cultures, institutions and materiality' that made up civil defence as a political and administrative project. In what follows, we focus on three cross-cutting themes that usefully highlight not only the interplay between imaginary and material and between representation and experience but also the highly differentiated ways the 'technology' of civil defence can be viewed when looked at from different perspectives. These themes are localities, temporalities and mediations.

Localities

Much work on sociotechnical imaginaries takes on an explicitly national-historical framework. Ulrike Felt argues that the concept:

> sensitizes us to how profoundly technologies are entangled with national technopolitical cultures [...] and how the (non-)development of specific technoscientific projects, on the one hand, and imagined preferred ways of living, value structures, and social order, on the other, are mutually constitutive. (Felt 2015: 104)

Felt focused on the peculiarly *national* response of Austria to new technology, but we can extend her insight both to a comparative dimension and to the wide range of various local contexts within nations. Civil defence, we can see, was understood differently within these local contexts, and our comparative national dimension makes this distinction all the sharper.

A clear pattern emerging from the answers in the Swedish questionnaire is the radically different experiences in rural areas compared to urban ones. Living in the sparsely populated countryside was still the reality for the absolute majority of Swedes in the 1950s and 1960s. 'You have to remember I lived in the countryside and I had no part in those things', says one woman, born on the Swedish west coast in the 1930s.[2]

Other informants also specifically address the lack of civil defence memories and experiences, referring explicitly to their rural upbringing. Nor did the much-discussed civil defence leaflet *Om kriget kommer: Vägledning för Sveriges medborgare* ('If the war comes: Instructions for Swedish citizens'), distributed to all households in Sweden in 1943, 1952 and 1961 (Cronqvist 2012), leave a trace. 'I have no recollections of total defence, shelters or evacuation rehearsals, and no memories of a civil defence leaflet', the same woman writes. The leaflet seems to have disappeared in the daily recycling of paper at the farm, as toilet paper in the outhouse or as material for lightning the fire. In fact, even the Cold War as a time period and concept seems to be quite vague with these informants. These rural postwar experiences indicate that maybe even the Cold War itself was considered an urban phenomenon.

In Sweden's more densely populated areas, the situation seems to have been very different. A typical answer here is the very detailed information from Gertrud, born in the city of Gothenburg, Sweden's second largest city: 'We had public shelters in the apartment complex where I lived, and I was a member of the civil defence and had to go through exercises in how you take care of injured people or the mentally challenged'.[3] Recollections from informants from later generations indicate similar experiences as parts of their everyday life. School evacuation drills were a common feature of being a child in Stockholm, Gothenburg or Malmö in the 1950s and 1960s, and the world's largest peacetime mass evacuation exercise involving a quarter of a million people was held in Stockholm in 1961. The construction of a series of mass public shelters for up to 10,000 people also defined the everyday cityscape in postwar urban Sweden (Cronqvist 2012, 2015).

The urban–rural divide was experienced very differently in the UK, mainly because civil defence was remarkably absent in urban centres for much of the Cold War (Grant 2011). There were no mass public shelters, and as the country was a central Cold War antagonist, it was perceived that, if war did come, British cities would simply be obliterated. This is perhaps the reason why voluntary civil defence struggled in British cities: a sense that it was fruitless. Real experiences of bombing during the Second World War could be one explanation, experiences that separate the UK from a country like neutral Sweden, which experienced no bombing at all. We can certainly see that urban civil defence became the focus of resistance to the idea that civil defence could work (the idea of civil defence as a technology, as it were). Opposition to civil defence focused on the idea that it was worthless, and Nicholas Barnett has written about a tense encounter between civil defence volunteers and local protesters in Coventry in 1954 (Barnett 2015). The fact that British cities would be targeted, and that civil defence could not protect people, helped entrench resistance to civil defence and promote the idea that it was a somewhat ridiculous endeavour.

In rural areas, however, voluntary civil defence was in a much healthier state. We can see three reasons for this. First, the volunteers were not in immediate target areas and therefore would be more receptive to the argument that although volunteers would have no impact in urban centres, much would be possible at the perimeter of the zone of destruction. Second, it could be argued that people in rural areas had different ideas of community and neighbourliness, which made them more likely to volunteer. Or, conversely, there was less competition for the leisure time of these people than there was in cities. Third, rural areas were also more conservative and contained more people who would seek leisure satisfaction within a uniformed, quasi-military government-sponsored service. And they were governed by elected local politicians who were much more pro-civil defence than their left-wing counterparts in the city.

One interview story neatly illustrates this divide. Robert volunteered for the Auxiliary Fire Service (AFS) in Cardiff, the major urban centre of Wales and, with its docks and industry, an obvious target. When asked about the usefulness of civil defence, he replied:

> People were talking about cities being obliterated and enormous fire storms and so on. And really, thinking back to it, the AFS wouldn't have been able to do very much. Not really. You could've contained the fire around the edge but there's no way you could've –[4]

At this point, Robert abruptly changed the subject, unwilling or unable to let his sentence reach this logical conclusion and admit the redundancy of civil defence. Instead, he went on to deliver a long and detailed anecdote about his AFS unit extinguishing a fire in Ludlow on the unit's way back from a training exercise. The story of defeating a barn fire allowed Robert to regain the composure lost when explaining the difficulties that the AFS would have faced in a nuclear war. It was a well-told story that had clearly been aired regularly. It also helped explain that the AFS was an effective organisation. After he had cast doubt on the usefulness of the AFS in the context of nuclear war, Robert's story highlighted circumstances where it, and his own service within it, had been effective and worthwhile. Later in the interview, Robert was able to return to the issue of usefulness with more confidence, arguing that 'I think we felt, yes, that we would be able to help, maybe perhaps only on the periphery of something, but we did have the equipment'.[5] Robert's Ludlow story might not be enough to convince critics of urban civil defence of its utility, but it allowed him to articulate that the voluntary services would have been of 'some' use by presenting a 'real-life' example of usefulness. This memory was a part of Robert's very positive experiences of civil defence. He concluded: '[W]e had a wonderful time. We were very late getting back to Cardiff but we, it was a great, a great Sunday'. In essence, civil defence in its rural setting was remembered positively as being active and manageable in a way urban civil defence could never be.

Temporalities

There are three interlinked aspects in the discussion of temporalities of civil defence memory in relation to the sociotechnical imaginaries framework: periodisation within the Cold War, a direct sense of being of a 'generation', and the unique impact the threat of nuclear war had on temporalities and the experience of time. Understandings of civil defence shifted over the course of the Cold War, as is to be expected. The interrelationship between the *technology* of civil defence and the political-military nature of the Cold War is vital to understanding the history of civil defence. Civil defence was viewed differently at different points in time, but it was also viewed differently depending on social class, gender and not least generation.

From the point of view of political history, whether the early fears were of a war fought with conventional weapons, atomic bombs or thermonuclear bombs, the civil defence imaginary had to adjust. Even in the thermonuclear era, civil defence was viewed differently in the 1960s, when 'mutually assured destruction' (MAD) was assumed, compared to the 1980s, when there were various debates about nuclear issues, including the possibilities of a 'limited war' and the prospect of a 'nuclear winter'. This changing *political* nature of the Cold War is key to understanding 'resistance' to civil defence. To put it simply, in the UK at least, there was increasing distance between what civil defence experts claimed the 'technology' could achieve (in terms of survival and protection) and what people believed would be the case (Grant 2010; Stafford 2012), to the extent that the state largely retreated from engaging in public education initiatives after the early 1980s (Preston 2015).

Resistance to the civil defence imaginary, then, could come in multiple forms and must be viewed differently across different periods *within* the Cold War. At its most basic, it could come in the form of denial—conscious or unconscious. One informant answering the Swedish questionnaire misunderstood the entire question about the Cold War, and instead answered by talking about the Second World War, explaining that '[a] big part of my childhood was marked by the Second World War'.[6] For him, the war was the important event, and what followed was less important. Another informant declared her standpoint by sending a very short reply to the questionnaire stating that she refused to answer any questions here because in her home as a child, one was not supposed to talk about 'the war'. However, it is notable that she did bother to send that short statement, as a way of 'talking back' to the questionnaire designer. Wally, a man who actually served in Civil Defence in the UK, was unable to recall any detail of his service, his motivations for joining or indeed the wider purpose of civil defence. He did, however, talk at length about both his experiences during the Second World War and his post-civil defence work volunteering with Army veterans.[7] Clearly for all these three voices, in their minds, the Second World War was the big event, both in terms of their own life course and emotional experiences and within wider cultural memory.

For others, however, the link between Cold War civil defence and the Second World War was a way of comprehending the subject, of making it more *knowable* and less abstract. In this sense, such memory work could be seen to form part of the process of *embedding*. One participant, Frank, served in the Auxiliary Fire Service arm of civil defence. He was born in 1948 and remembered the wartime generation: 'We were fortunate enough to have guys who'd been firefighters. Our gaffer, uh, John Reynard, he'd been a firefighter in the war anyway'. Central to this was the sense of expertise of an experienced generation, 'they carried on in the Auxiliary Fire Service, so we had good people training us, you know, they'd pass their knowledge on'.[8] Considering all the fundamental differences between Sweden and the UK in the experience of the Second World War and not least of the bombing of civilian targets, the extent to which the war constantly functions as a point of reference is as striking in Sweden as it is in the UK. For a certain generation, it seems to be impossible to even talk about the Cold War as something other than a prolongation of the Second World War.

To return to the subject of what in the sociotechnical imaginary framework is named *resistance*, in our material, political opposition to civil defence is also visible. But it could also take the form of objecting to civil defence's coercive nature in general and its place within governmental state authority. A vital part of the civil defence 'technology' was its position within the state, and states having the ability to control their population were essential to the idea that civil defence could indeed 'work' and supply a (more) desirable future. Karin, a Swedish nurse born in 1929, remembered: 'When I was 15 or 16, me and my classmates were called by the civil defence to a cottage filled with tear gas. You were supposed to run through that cottage. One of my friends was pregnant and refused to show up. She was collected by the police and was forced to run through the cottage'.[9] In the UK, much of the resistance to civil defence was along similar lines: that it was a symbol of state power. It is no surprise that many of the most memorable scenes from popular film and television concentrated on this theme, including the police shooting looters in *The War Game* (1965) and armed traffic wardens in *Threads* (1984). In the same vein, 'bunkers' were the preserve of the authorities and denied to the public.

One of the constants of Cold War historiography is the debate about when it started, when it ended and whether it existed at all (see Stephanson 2012; Nehring 2012; Romero 2014). In many ways, the 'Cold War' only becomes a coherent historical event or period in retrospect. Even then, the differences between considering it an event (a 'war') and a period (in which all events and experiences are understood as occurring within the 'Cold War Age', as it were) are elided when civil defence is made the focus of study. Civil defence was both a clear part of the fabric of the Cold War-as-war *and* part of the fabric of everyday life. This is why civil defence is so important to study as the point at which the military conflict enters the realm of ordinary experience. A good example of this comes from a male informant, born in the south of Sweden, who said: 'I was born in 1960 and am one of millions of people in what we can call the "Cold War generation"'. However, as he goes on remembering his childhood, he is in a paradoxical way reluctant to relate this generational experience to real Cold War fears. 'I don't think that, living in Sweden, we thought very much about the world being on the brink of a nuclear war', he says.[10] This is highly revealing of Cold War memory: a general sense of living though the 'Cold War', being part of that 'generation', but struggling to pin down any specific experiences or immediate relevance. We could posit that this is typical of respondents who want to 'forget' the fear of the Cold War, or who are 'unable' to remember it, but it could equally be suggested that this complex and difficult memory work is shaped exactly by the effect of civil defence. Viewed in terms of a sociotechnical imaginary, the memory of a Cold War generation highlights the 'pastness' of civil defence and its relative lack of purchase in contemporary life.

Within memory work, however, those recalling civil defence and the threat of nuclear war are re-encountering a past which, for many, had no future. To put it starkly, the Cold War involved large numbers of people imagining that there *was no future*, that historical time was not an open horizon (Grant 2016). The social, cultural and psychological impact of this imaginative labour has yet to be fully explored, yet the histories of civil defence, the global peace movements, and political and diplomatic notions of nuclear deterrence are scarcely possible without the fundamental notion that the end of humanity, of time, had to be confronted (see Andersson 2012; Connelly et al. 2012). Of course, civil defence was both an attempt to avoid such an apocalyptic catastrophe, instead imagining a more desirable future, and a way of rationalising it. But the fact that individuals struggle to adequately remember this process, to bridge that temporal gap, highlights the need to understand the civil defence imaginary in more nuanced, textured ways. In this sense, it is perhaps useful to bear in mind that within Cold War generations, memory work involves the participants attempting to reconstruct different 'planes of historicity' (Koselleck 2004: 9–25), and one of the contributions a study of civil defence can make to rethinking sociotechnical imaginaries as a field of analysis is to provide a more nuanced questioning of how 'the future' part of 'desirable futures' was understood.

Central to this aspect of memory work is the remembering of ways civil defence was rationalised and how people *coped* with the threat of nuclear wars. A number of the Swedish informants, especially those born in the 1920s and 1930s, remember their teenage years as a period of non-engagement in politics and global issues. One aspect of this was preoccupation with the 'everyday', with growing up, child-rearing and other aspects of the intimate sphere, all of which kept people's minds busy and relegated foreign policy to the fringes of concern. But the comments also reveal a form of critique of the Cold War mindset imposed from above. Commenting on the civil defence leaflet 'If War Comes', one person remembers: 'I kept [it] in a special place. But I guess it was mostly to show my children the value of keeping things in order at home'.[11]

Memories of child experiences of evacuation drills are also visible in the answers to the Swedish questionnaire, and often these memories are narrativised precisely along the lines of what Jasanoff and Kim call resistance or a form of distancing from the bizarre daily life of adults. One woman, born in 1932, remembers watching the adults evacuating. 'As children we stood and giggled when the fat ladies ran. We found it hilarious', she remembers.[12] Another informant, born in 1936, is even more explicit in his criticism. He remembers civil defence in the 1950s as something he and his friends considered more as a joke than a serious matter. '"Wash yourselves in ordinary soap and water" seems to have been the standard answer to any question', he remembers, and he also recalls that he read civil defence brochures 'without any enthusiasm'. The Cold War did not have much effect on his life or feelings at all, he concludes.[13]

In a similar way, three women were interviewed as a group about their experiences working in Coventry's Civil Defence Headquarters in the 1950s and 1960s. They had very different memories of the nuclear threat and expressed very different emotional responses to questions, something that threatened to disrupt the interview entirely. The emotional tenor of the interview, and the inter-subjective encounter, was 'saved' by discussing the small details of daily office life, especially its more domestic aspects: the coffee breaks, the co-workers, even the large bath which was available for them to use.[14] The interview encounter asked them to place themselves at a certain moment of historical time, and to attempt to recover their emotional responses. This is enormously difficult, not to say impossible. The fact that it led to quotidian memories of the everyday is far from meaningless in Cold War terms, however. It highlighted the role of such memories in securing people in narratives of the past which were more recoverable, and more comforting, than discussing the nuclear threat.

Mediations

Cold War civil defence memory work, it could be argued, is particularly open to the entanglement of popular culture in people's recollections. As it was experienced largely as an imaginary, it naturally meant that people relied on existing narratives and images to make sense of it at the time, and to make sense of their memories of it subsequently. The lack of cultural value attributed to civil defence in the years following the Cold War would have only deepened the sense that it was hard to remember. Put simply, the mediated nature of memories highlights that the memory of civil defence is at the same time creating and reflecting cultural expressions. This means that any hard divide between ideas/materiality or even structure/agency are impossible or perhaps even meaningless. We argue that the attention to mediated memories problematises, or even dissolves, potential hard binaries in the sociotechnical imaginaries framework—a dissolution in fact also encouraged by Jasanoff herself in the introduction to the book *Dreamscapes of modernity* (Jasanoff and Kim 2015: 29).

It is clear from our oral history sources both the answers to the Swedish questionnaire and the UK interviews—that what people seem to remember about civil defence is often deeply entangled in, supported by and sometimes contradicted by media images, news articles and popular culture. In this way, cultural memory is always already mediated—or even remediated (Erll and Rigney 2009). Therefore, it is fruitless and ahistorical to attempt to completely disaggregate the social experience of the Cold War from popular culture within individual memory; both are indeed central to memory work. Again, this reminds us of the 'fuzziness' of both memory and everyday life, and that perhaps the extension–resistance–embedding processes of the sociotechnical imaginaries framework need to be seen as more fluid than the language of 'turning imagination into social practice' suggests, perhaps even as anti-linear (Jasanoff and Kim 2015: 323).

Media images are constantly present on different levels in the Swedish material. References to and memories of civil defence and the Cold War in general are often surrounded by recollections of everyday media practices and iconic historic events told by or channelled through the media. One informant remembers sitting by the radio with his father one night, listening to the first atom bomb test at Bikini atoll. Others simply remember parents reading newspapers, listening to radio, watching television and talking with low voices about Korea, Cuba or Vietnam.[15] Margareta from Stockholm recalls evacuation exercises and civil defence drills of the early 1960s; however, big Cold War events were mediated in her everyday life in a completely different way. '[T]here was a hero from the Soviet Union, Gagarin who was an astronaut [sic]', she notes. 'What an exciting man he was, and handsome too. I put up his picture on the wall in my room'.[16] Another informant, Ingrid, remembers the immediate postwar years as filled with exciting American experiences: chewing gum, Coca-Cola, nylon stockings, movie stars and 'the new look'.[17] The generational aspect is also dominant here, as media images of the Cold War as an everyday experience shift. Ida and Nils, both born about twenty years later in the 1970s, mainly associate the Cold War with sport events such as the 1980 and 1984 Olympic Games in Moscow and Los Angeles or the iconic World Chess Championships in Reykjavik and elsewhere. '[I]t was easier to see the conflict as a hockey game', Ida writes. She also remembers a nuclear shelter at her school, but it was used as a rehearsal studio for her rock band—enforcing perhaps the 'underground' feeling of engaging with youth and subculture in a military space.[18]

Even among the UK volunteers who served in civil defence, mediated images were central to their understanding of the nuclear threat and the ability of the civil defence organisation to tackle it. Julian, a volunteer from Cardiff, remembered 'it was interesting and frightening [...] when you hear about the effect these things had. I mean, you had these films which showed the effect of an atomic blast, where you can see the buildings and then the dust and then the wind and the buildings gone'.[19] Crucially, the filmed images of the bomb were central to the imaginative gameplay that typified civil defence. 'And we did an exercise when we were up at Easingwold [the civil defence training centre in the North of England] and they dropped a five-kilotonne bomb on the centre of Reading. How do you deal with it? Which is what we then did. And as I say, there was one bloke who reckoned we should forget about the centre because they'd be gone. And concentrate on the outer part. All the rest of us weren't willing to abandon the people in the middle you know, but having said that, they were probably all dead anyway'.[20]

One of the interviews was a complex exchange involving three women who worked in the civil defence headquarters of Coventry. They were not volunteers, but paid employees of the local council. They remembered the training films for atomic blasts. Pamela, who left the council to join the police force, remembered:

> the police cadets used to come up and borrow them for their training didn't they, before I joined the police force, and they used to take them back to police headquarters and then bring them back; I can remember they were in a filing cabinet at the end of the office [...]. But then when I was a cadet, I had to watch those films in the basement of police headquarters. Police headquarters in Coventry has got um, a, supposedly,

> bomb-proof basement [...]. Yeah, so we used to watch those films down there. Absolutely terrifying.[21]

The discussion of the films led one participant to reveal the depths of her anxiety about nuclear war. Doris said:

> The thing I remember mostly about this was I used to get quite worried about the atom bombing. I remember being quite concerned about my parents, what would happen to them because we would go into the air raid shelter, the control place, and my parents would be on their own. And uh, I used to try and persuade people to [...] where I lived it was sort of square houses with great long back gardens, and I used to try to persuade people to build a shelter in the middle of this and fit it out with food, all the things we were supposed to do [...]. I used to try very hard to get them to build a shelter so they could all go into it, but nobody would listen to me.[22]

The third member of the group, Barbara, had very a very different memory: 'No, I never saw that training film you were on about, I can't ever really have been worried about anything'.[23]

At the level of everyday practice, the civil defence organisation relied on imagination, providing a simulacrum of defence against nuclear war. Images of nuclear war were just as central to civil defence activists as they were for the mass of 'ordinary' people who confronted the Cold War. Lars Nowak has written about the centrality of training films for civil defence in the United States (Nowak 2016), and in the UK, representations of nuclear war were central in ways that illustrate the interrelationship between the imaginary and the material (see Chapter 5 by Bennesved and Sylvest in this volume). Civil defence exercises remind us that civil defence was performative, and that this performative element was intrinsically material. Civil defence activities often involved physical reconstructions of the post-attack city used to demonstrate how civil defence would save lives (Grant 2010). Moreover, Cold War civil defence could be read onto the materiality of urban space, as Coventry was still in the midst of a slow process of reconstruction after the devastating attack suffered during the Second World War. Memory (of a different war), imagination and materiality could not be separated. Jessica Douthwaite, in an important article on the complexity of the memory work undertaken by former civil defence volunteers who attempted to reconstruct how training exercises 'rehearsed' nuclear war, also argues that mediation and materiality were inseparable. As the rehearsal of a 'deferred' nuclear attack was remembered, Douthwaite's participants were able to reach 'a historical understanding of mediated consciousness' (Douthwaite 2019: 199).

Conclusion

A key aspect of engaging with the sociotechnical imaginaries framework is investigating the concept's usability. What are the strengths and weaknesses of the concept when transferred to other fields of analysis? The aim of this chapter has been to explore how the framework can be used to examine a technology that is remembered in a fundamentally altered political-cultural context. The chapter has employed a historical ethnography approach, using primarily oral history sources, interviews and questionnaires collected between 2006 and 2012. We have chosen a comparative approach with examples from the UK and Sweden, revealing both similarities and fundamental differences between the two countries. But also—and perhaps more importantly—we have shown, through our empirical examples from the UK and Sweden respectively, that people's memories tend to constantly diverge from the neat paths of the civil defence imaginary. Essentially, there was no single way of understanding civil defence at the time and certainly no single way of remembering it decades later. Understandings were, and still are, shaped and reshaped by social and cultural factors such as locations, temporalities and mediations.

But the evidence also highlights that the interlinking of the imaginary and the material led to civil defence being at the same time curiously unrooted in memory and (Cold War) wider society. The chapter serves, then, as a cultural history of the circular feedback-loop implied in the *origins–embedding–extension–resistance* processes. Our case studies highlight some of the difficulties people had in fully grasping civil defence or fully recapturing its meaning. When analysing the course of a sociotechnical imaginary through cultures and societies, more attention has to be paid to the quotidian, messy and partial ways it was understood and sometimes rejected in everyday life.

Our chapter also illustrates that the emotionology of civil defence could not be fully grasped through traditional analyses of top-down 'emotional management' but requires complementary 'bottom-up' approaches. This cognitive gap between emotion and memory in civil defence history needs further exploration, and we hope the comparison in this chapter of oral histories in Sweden and the UK has served to open up such a discussion.

It could be a function of the operation of memory mentioned above. It could provide clues about the everyday manoeuvring required to adjust to civil defence, to confront the anxiety seemingly inherent in the Cold War nuclear confrontation and to reason out debates about nuclear security, deterrence, protest and disarmament. The 'emotional navigation' charted by individuals may have been as much about *not* engaging with nuclear anxiety as anything else, but civil defence as a sociotechnical imaginary certainly asked people to confront that choice even as it sought to provide reassurance. The non-engagement, the refusal to answer questions about civil defence, the complete lack of memories and the misunderstandings—conscious or unconscious—are sometimes also paradoxical statements in themselves. They could at the same time actively resist and confirm the threat of nuclear war. In the words of Carl-Erik, born in a small Swedish town in the early 1930s: 'My daily life really didn't change. We humans do have a remarkable ability to blank out the unbearable'.[24]

The complexity and divergence of everyday life are as familiar to any ethnographer as the multilayered processes of remembering are known to the researcher of historical culture. As the vast and significant scholarship on memory studies keeps reminding us: memories are not merely experienced, but are constantly made and remade in the light of new presents. What this chapter has brought to the table is a discussion of how we may complement or refine the sociotechnical imaginary framework to incorporate at least some elements of the fuzziness of everyday life and how these elements of everyday culture relate to processes of embedding, resistance and extension of civil defence in Sweden, the UK and beyond.

Notes

1. LUF220 Kalla kriget, Folklivsarkivet (LUF), Lund. The questionnaire was initiated by Marie Cronqvist as a part of the research project 'The people's home in the atomic age: Civil defence and the Swedish narrative of community', financed by the Swedish National Research Council (VR) 2005–2009.
2. LUF 220: Astrid, b. 1937.
3. LUF 220: Gertrud, b. 1927.
4. Interview with Julian and Robert.
5. Interview with Julian and Robert.
6. LUF 220 Kalla kriget. Bengt, b. 1935.
7. Interview with Wally.

8. Interview with Frank.
9. LUF 220 Kalla kriget. Karin, b. 1929.
10. LUF 220 Kalla kriget. Staffan, b. 1960.
11. LUF 220 Kalla kriget. Harald, b. 1921.
12. LUF 220 Kalla kriget. Lilian, b. 1932.
13. LUF 220 Kalla kriget. Thorsten, b. 1936.
14. Interview with Barbara, Doris, and Pamela.
15. LUF 220 Kalla kriget. For example Rune, b. 1933; Gertrud, b. 1927, Ingrid, b. 1934.
16. LUF 220 Kalla kriget. Margareta, b. 1950.
17. LUF 220 Kalla kriget. Ingrid, b. 1934.
18. LUF 220 Kalla kriget. Ida, b. 1974.
19. Interview with Julian and Robert.
20. Interview with Julian and Robert.
21. Interview with Barbara, Doris and Pamela.
22. Interview with Barbara, Doris and Pamela.
23. Interview with Barbara, Doris and Pamela.
24. LUF 220 Kalla kriget. Carl Erik, b. 1931.

References

Unpublished Sources

The Folk Life Archives (Folklivsarkivet; LUF) with the Scania Music Collection (Folklivsarkivet med Skånes musiksamlingar), Questionnaire no. LUF 220 'Kalla kriget'.
Interviews with former civil defence officials in the UK (recordings and transcripts).

Literature

Abrams, Lynn. 2010. *Oral history theory.* London: Routledge.
Andersson, Jenny. 2012. The great future debate and the struggle for the world. *American Historical Review* 117: 1411–1430.
Assman, Aleida. 2011. *Cultural memory and Western civilization: Functions, media, archives.* Cambridge: Cambridge University Press.
Barnett, Nicholas. 2015. 'No protection against the H-Bomb': Press and popular reactions to the Coventry civil defence controversy, 1954. *Cold War History* 15: 277–300.

Confino, Alon. 1997. Collective memory and cultural history: Problems of method. *American Historical Review* 102: 1386–1403.
Connelly, Matthew, et al. 2012. General, I have fought just as many nuclear wars as you have. *American Historical Review* 117: 1431–1460.
Cronqvist, Marie. 2012. Survival in the welfare cocoon: The culture of civil defence in Cold War Sweden. In *Cold War cultures: Perspectives on Eastern and Western societies*, eds Anette Vowinckel, Marcus M. Payk and Thomas Lindenberger, 191–210. New York: Berghahn Books.
Cronqvist, Marie. 2015. Evacuation as welfare ritual: Cold War media and the Swedish culture of civil defence. In *Nordic Cold War cultures: Ideological promotion, public reception, and East-West interactions*, eds Valur Ingimundarson and Rosa Magnusdottir, 75–95. Helsinki: Aleksanteri Cold War Series.
Dawson, Graham. 1994. *Soldier heroes: British adventure, empire and the imagining of masculinities*. London: Routledge.
Derrida, Jacques. 1986. *Memories for Paul de Man*. New York: Columbia University Press.
Douthwaite, Jessica. 2019. '...what in the hell's this?' Rehearsing nuclear war in Britain's civil defence corps. *Contemporary British History* 33: 187–207.
Erll, Astrid, and Ann Rigney, eds. 2009. *Mediation, remediation and the dynamics of cultural memory*. Berlin: Walter de Gruyter.
Felt, Ulrike. 2015. Keeping technologies out: Sociotechnical imaginaries and the formation of Austria's technopolitical identity. In *Dreamscapes of modernity: Sociotechnical imaginaries and the fabrication of power*, eds Sheila Jasanoff and Sang-Huyn Kim, 103–125. Chicago: Chicago University Press.
Grant, Matthew. 2010. *After the bomb: Civil defence and nuclear war in Britain, 1945–1968*. Basingstoke: Palgrave Macmillan.
Grant, Matthew. 2011. 'Civil defence gives meaning to your leisure': Citizenship, participation, and cultural change in Cold War recruitment propaganda. *Twentieth Century British History* 22: 52–78.
Grant, Matthew. 2016. The imaginative landscape of nuclear war in Britain 1945–65. In *Understanding the imaginary war: Culture, thought and nuclear conflict, 1945–90*, eds Matthew Grant and Benjamin Ziemann, 92–115. Manchester: Manchester University Press.
Grant, Matthew. 2019. Making sense of nuclear war: Narratives of voluntary civil defence and the memory of Britain's Cold War. *Social History* 44: 229–254.
Jasanoff, Sheila. 2015. Imagined and invented worlds. In *Dreamscapes of modernity: Sociotechnical imaginaries and the fabrication of power*, eds Sheila Jasanoff and Sang-Huyn Kim, 321–341. Chicago: Chicago University Press.
Jasanoff, Sheila, and Sang-Huyn Kim, eds. 2015. *Dreamscapes of modernity: Sociotechnical imaginaries and the fabrication of power*. Chicago: Chicago University Press.

Koselleck, Reinhart. 2004. *Futures past: On the semantics of historical time*. New York: Columbia University Press.

Nehring, Holger. 2012. What was the Cold War? *English Historical Review* 127: 920–949.

Nowak, Lars. 2016. Images of nuclear war in US government films from the early Cold War. In *Understanding the Imaginary War: Culture, thought and nuclear conflict, 1945–90*, eds Matthew Grant and Benjamin Ziemann, 260–86. Manchester: Manchester University Press.

Oakes, Guy. 2004. *The imaginary war: Civil defence and American Cold War culture*. Oxford University Press.

Passerini, Luisa. 2003, Memories between silence and oblivion. In *Contested pasts: The politics of memory*, eds Katherine Hodgkin and Susannah Radstone, 238–254. New York: Routledge.

Preston, John. 2015. The strange death of civil defence education. *History of Education* 44: 225–242.

Reddy, William M. 2001. *The navigation of feeling: A framework for the history of emotions*. New York: Cambridge University Press.

Romero, Federico. 2014. Cold War historiography at the crossroads. *Cold War History* 14: 685–703.

Rosenwein, Barbara H. 2006. *Emotional communities in the early Middle Ages*. Ithaca, NY: Cornell University Press.

Stafford, James. 2012. 'Stay at home': The politics of nuclear civil defence, 1968–83. *Twentieth Century British History* 23: 383–407.

Stephanson, Anders. 2012. Cold War degree zero. In *Uncertain empire: American history and the idea of the Cold War*, eds Joel Isaac and Duncan Bell, 19–49. New York: Oxford University Press.

Summerfield, Penny. 1998. *Reconstructing women's wartime lives*. Manchester: Manchester University Press.

Summerfield, Penny. 2004. Culture and composure: Creating narratives of the gendered self in oral history interviews. *Cultural and Social History* 1: 65–93.

CHAPTER 10

Conclusion: Civil Defence Futures (Re)imagined

Marie Cronqvist, Rosanna Farbøl, and Casper Sylvest

As a collective effort, *Cold War Civil Defence in Western Europe* has sought to bring together new and established perspectives in civil defence history within a theoretical framework sensitive to transnational dynamics: from mighty international organisations to local groups in peripheral areas, from top-down plans and ambitions of authorities to everyday practices of individuals on the ground, from futuristic visions to fuzzy memories and hard materialities, from embracement to resistance. As a whole, the book highlights important and extensive European variations in how societies and peoples grappled with the nuclear threat and how

M. Cronqvist
Department of Communication and Media, Lund University, Lund, Sweden
e-mail: marie.cronqvist@kom.lu.se

R. Farbøl · C. Sylvest (✉)
Department of History, University of Southern Denmark, Odense, Denmark
e-mail: csy@sdu.dk

R. Farbøl
e-mail: rosfa@sdu.dk

M. Cronqvist et al. (eds.), *Cold War Civil Defence in Western Europe*,
https://doi.org/10.1007/978-3-030-84281-9_10

they were transformed in the process, but it also points to similarities, as well as a myriad of transnational connections. Together, the chapters enrich our understanding of civil defence in the nuclear age within and across national boundaries. In this short conclusion, we offer a few reflections on three themes. First, we bring up the value and challenges of the theoretical approach employed in this book. Second, we highlight the most striking findings, commonalities and differences across the chapters. As a third and final point, we stress the importance of civil defence history at a time when policymakers are once again seeking to promote preparedness and devoting energy and resources to its embedding in cultures, institutions and citizenries.

Theoretical Reflections

One starting point for this book was to explore the merits and potential of studying civil defence through the concept of sociotechnical imaginaries (Jasanoff and Kim 2015; Jasanoff 2015a, b). This approach, presented and discussed in the Introduction to the volume, has provided a common vocabulary, which was central to the ambition of moving beyond a decidedly national tilt in civil defence history and making linkages visible and comparisons possible. Indeed, the framework has sharpened our analytical focus and facilitated comparisons, while allowing for diversity, dynamism and creativity. Moreover, it has allowed us to highlight the central role of technology in civil defence and to study the multiple forms of power that were involved in civil defence practices, whether they involved the rehearsal of strategies for survival during catastrophes, the disciplining of citizens or the (re)shaping of societies and everyday life. While technology and technological development has been a subtext of several existing studies of civil defence history, the benefits of working with a common vocabulary that explicitly theorises the role of technology in social life has allowed us to identify ways in which both nuclear and non-nuclear technologies were integral to civil defence as vision and practice. Finally, the concept of sociotechnical imaginaries has cast light on the social embeddedness of civil defence by focusing attention on how this concept prompted new forms of social interaction and reflected existing social norms in an age when nuclear technologies were swathed in fear of doom and hope of survival. The depth and breadth of the chapters testify to these advantages.

To be sure, the versatility of the concept of sociotechnical imaginaries—its capacity to be harnessed for a wide variety of research questions and themes in the human sciences—goes a long way towards explaining its uptake and the increasing popularity of the line of questioning the wider approach entails. Yet, this is not the entire story. Another attraction of this approach lies in its tendency to incorporate and connect to a wide range of themes and lines of enquiry that are central to contemporary research in the humanities and social sciences. They include, self-evidently, a resurgence of interest pertaining to the role of science and technology in human life but also a curiosity about the human faculty of imagination, the role of futures in social and political life, a sensitivity to the multifaceted forms of power that are enacted in these contexts and a renewed appreciation of the role of materiality in human interaction.

That said, it is a striking feature of the growing literature that it focuses on brighter visions of technological possibilities—or, to be precise, visions that, although they can be (and are) criticised, centre plainly on the benefits rather than threats to human beings and the world they inhabit. Examples of these imaginaries range from autonomous driving and energy transition to digital transformation, genetically engineered crops and the fourth industrial revolution.[1] Many of these studies highlight the generative potential of technological future thinking whereby actors not only formulate visions but also (seek to) 'perform and produce' them (Urry 2016: 9). Such studies also harness the critical potential of the concept of sociotechnical imaginaries by highlighting the inconsistencies, gaps or flight of fancy that they tend to entail, thus making clear that sociotechnical imaginaries are 'contested, changeable, flexible and loose around the edges' (Sismondi 2020: 505).

Still, it is remarkable that few studies have so far employed the concept for those imaginaries that are negative, protective or preserving in nature—imaginaries that are overwhelmingly found in areas of health, safety, security and risk. Civil defence is one such case, focused as it is on a safeguarding vision of ensuring resilience, the continuation of social life or bare survival in the face of total war. Overall, we think the chapters in this book demonstrate that the concept of sociotechnical imaginaries is useful for the study of an inherently ambiguous imaginary like civil defence, caught as it was between dystopian and utopian projections.

Fulfilling the ambitions of this collective project and operationalising the theoretical approach in diverse historical contexts and at various levels of analysis has, at times, been complicated. We set out with an open mind to explore the potential of the concept of sociotechnical imaginaries in empirical, historical research, and at an early stage we identified a series of challenges, some of which are dealt with briefly in the Introduction to this volume. In addition, several chapters raise and engage with potential shortcomings of the approach such as the tendency to foreground top-down perspectives and the inclination to approach imaginaries within a linear, perhaps even teleological, perspective at the expense of the complexity that historians traditionally emphasise. What emerges from the attempt to use the approach in historical studies of civil defence is the importance of perspective and context for assessing the ways in which imaginaries functioned. This comes to the fore in the chapters that focus on the public sphere or on how civil defence was practised on the ground.

In exploring the concept of sociotechnical imaginaries empirically, the chapters demonstrate that it can indeed be adjusted to inform a range of analyses, just as it is compatible with a series of perspectives and priorities that emerge from the humanistic interest in human beings as both the producers and products of history. Indeed, the contributors to this volume have developed important theoretical adjustments and refinements, while also adding historical depth to the concept of sociotechnical imaginaries. Jasanoff's definition can be read as an instinctively positive evaluation of the role of science and technology in that complex historical development we conveniently (but loosely) refer to as modernity. One example of this is the emphasis on 'desirable futures' in the definition of these imaginaries and their propensity to be 'supportive of' advances in science and technology. On closer inspection, however, and as Bjørnsson discusses in Chapter 2, the concept allows for a wider application and can accommodate a greater degree of complexity than the definition may at first glance suggest. In the context of civil defence, the desirable future is seldom a bright future. The desirable is to find a way to collectively handle the demanding threats to society.

Applying the concept to civil defence places special emphasis on the series of ambivalences about science, technology and progress that often accompanies visions of human futures and modernisation. In some cases, these traits are not only found in forms of resistance but also exist at the very heart of imaginaries. In that context, it is also worth stressing how the notion of civil defence as (forming part of) a sociotechnical imaginary constitutes a balancing act by appealing at once to an amalgam of the ideals of a security paradigm (technologised efficiency, organisation, preparation) and a complex of existing norms and values (ranging from social beliefs to humanitarian values). Crucially, these commitments are juggled against a backdrop of uncertainty and controversy created by unfathomable weapons technologies. Theoretically, this points to the importance of legitimacy and credibility in the study of imaginaries revolving around controversial technologies that carry high risks. These themes deserve more theoretical attention within the study of sociotechnical imaginaries.

The intricacies of human–technology relations are revealed especially in the context of the Cold War and in the shadow of nuclear annihilation. Using a wide array of historical sources and analytical approaches, the book advances a multifaceted understanding and application of the concept of sociotechnical imaginaries that highlights social and historical processes in all their ambiguity and 'fuzziness', as Cronqvist and Grant (Chapter 9) term it. Sociotechnical imaginaries of civil defence were real and affected European societies and the lives of people. They were much more than plans, communication, initiatives and constructions rolled out from the top of those societies or their central institutions. While a range of chapters in the book highlight and study these 'official' imaginaries and the myriad attempts at embedding them in norms, institutions and ways of thinking, it is a distinct contribution of this volume to the theoretical discussion of this approach that it underscores how sociotechnical imaginaries are also produced, transformed and inhabited by individuals, organisations and practices in local, regional or professional contexts. We see such processes at work in the social life of North-Western England of the 1950s (Chapter 4), in the development of German disaster medicine (Chapter 3), in the ruin towns of provincial Denmark (Chapter 7) and in opposition to civil defence in the Netherlands and Switzerland (Chapters 6 and 8). This everyday and bottom-up perspective is vital for understanding how sociotechnical imaginaries function.

Common Themes and New Insights

The book has explored the nature and infrastructures of civil defence imaginaries at the international, transnational, national, regional and local levels. Across these boundaries, a common feature in several of the studies collected in this volume is the Western European delay or difficulty in adapting to the nuclear age. Well into the 1950s, the atomic bomb seemed more like an abstraction compared to the more immediate experience of suffering in the recent war. Hence, early imaginaries of civil defence looked to the European past to see the future. The only wartime nuclear attacks the world had experienced, those on Hiroshima and Nagasaki in August 1945, played a surprisingly marginal role. On the European continent, in contrast to a good deal of American civil defence thinking, which saw the beginnings of a transition towards atomic war already in the late 1940s and early 1950s, it was the recent destruction of German cities like Hamburg, Dresden and Cologne and the air raids on allied cities like London that constituted the framework for understanding total war and nuclear defence.

The importance of the Second World War in the early development of European civil defence is highlighted in most of the chapters in this book. Referring to what he calls 'the underlying imagined disastrous', Jochen Molitor states in his contribution (Chapter 3) that this 'was not so much based on the theoretical premises of nuclear warfare, but rather on first-hand experiences of the allied bombings of the recent past'. Furthermore, as demonstrated by Cronqvist and Grant (Chapter 9), for some civil defenders, memories of the Second World War trumped understandings of the Cold War—even when they were explicitly asked to talk about the latter. References to Hiroshima were not absent, but they often served the function of highlighting the astonishing and increasing yield of atomic bombs and confirming a depressing trajectory of modern warfare. The atomic bomb was overwhelmingly shoehorned into an existing world view, despite the recognition, hesitantly at times, that the new weapons technology presented a new, but difficult-to-grasp, invisible danger: radioactivity.

It was only with the arrival of the H bomb during the mid-1950s that a more systematic rethinking—especially in the public sphere—of the nature of nuclear war and, consequently, sociotechnical imaginaries of civil defence that were markedly nuclear in orientation really emerged (see in particular Chapters 2 and 5). The centre of gravity now lay in an unknown and unstable future. In the House of Commons in March 1955, Winston Churchill famously argued that the British people had entered a period in which safety would be 'the sturdy child of terror and survival the twin brother of annihilation' (Bourke 2006: 261). What this precisely meant for civil defence was not clear at the time, but its effects in Western Europe came to display both similarities and differences.

One common feature of civil defence thinking in the wake of the H bomb was that atomic or nuclear war was, on the one hand, accorded more explicit attention in civil defence thinking, communication and practice, while, on the other hand, often appearing as an un(der)specified phenomenon. This was the case, for example, in Britain, Germany, Denmark and Sweden, as illustrated by Molitor (Chapter 3), Hogg (Chapter 4), Farbøl (Chapter 7), and Bennesved and Sylvest (Chapter 5). Strikingly, these countries are marked by obvious and important differences: a nuclear-armed NATO member state (the UK), a NATO member state hosting foreign nuclear weapons (West Germany) and two non-nuclear states (Denmark and Sweden), one a NATO member and the other following a policy of armed neutrality (while secretly pursuing a nuclear weapons capability). Despite these differences, it was a marked feature of the sociotechnical imaginaries of civil defence that were created in these countries, above all by authorities, that there were few, if any, references to the causes or specific nature of the kind of catastrophe that could occur.

At the same time, however, mediations in many of the countries often legitimised ideals of preparedness by referring to (or even exploiting) an abstract sense of *Angst* and unease. Concrete civil defence practices and exercises also tended to operate with less catastrophic scenarios than full-scale nuclear war. In her chapter on Danish civil defence ruin towns, Farbøl points out that a key line of reasoning was that there was no point in preparing for the impossible. Instead, resources were geared towards manageable scenarios. These include conventional war and limited nuclear attacks (see Chapters 4, 5 and 7), but also peacetime nuclear accidents (Chapters 3 and 8), natural disasters and traffic accidents (see, for instance, Chapters 3 and 6), which paved the way for the kinds of all-hazards approach to emergency management prevalent today. Whether this logic is inherent to imaginaries—that their proponents create meaning by representing their shared visions as seamless and coherent—is an open question.

This recurring calibration of sociotechnical imaginaries of civil defence that sought to strike a balance between capabilities and ambitions and attune new visions of preparedness to existing social norms, expectations and practices is a striking, cross-cutting theme that emerges from this volume. The complex of norms and practices that such imaginaries had to accommodate included not only mundane habits of everyday life but also (gradually shifting) gender roles, notions of good citizenship, and national narratives and identities related to security policy or the welfare state.

In repeatedly performing this calibration, also in public, the risk of overstretch was always present, however. Once cracks appeared in the armour of civil defence, and they inevitably did, a host of other questions about its moral justification or allegations of its being out of touch with reality gave rise to (self-)criticism and resistance. In the wider public sphere, and especially among sceptics, detailed speculations about the political dynamics behind a nuclear war and extrapolations of the consequences of war for human life based on scientific opinion served to destabilise official imaginaries. When civil defence demonstrated its efficiency, it was liable to criticism that it was not concerned with the most important threats, and once the limitations to its value in nuclear war were recognised, it was vulnerable to charges of uselessness. Resistance to a sociotechnical imaginary could take the form of friction or outright opposition. In Chapter 6, van Lente details how a maternalist, grassroots imaginary of the women's civil defence organisation developed partly in opposition to the official, paternalist imaginary. At the other end of the spectrum, Molitor (Chapter 3) and in particular Marti (Chapter 8) demonstrate how civil defence was sometimes met with scepticism and even outright resistance, because it was seen as futile or worse: as contributing to the arms race and the likelihood of nuclear war.

The idea that civil defence itself constituted a militarisation of society was a refrain of peace movements in Western Europe—especially during the 1980s—and scholarly research, particularly in the USA, has detailed how everyday life was affected by the logic of militarisation throughout the Cold War (McEnaney 2000). Many mundane aspects of life—from family relations to workplaces and infrastructures of living and mobility—were transformed and assigned new, additional functions in light of the ever-present risk of nuclear war. Dual-use shelters, for example, crept into the foundations of private and public life. Yet, a larger question that emerges from this book is whether the militarisation thesis puts the cart in front of the horse. Several analyses in this book point in the opposite direction, namely that the increasing public circulation of ideas about survival and preparedness amounted instead to an 'everydayification' of the military.

In the era of total war, the home front mattered more than ever, but the war also took on a new quality as an ever-present, everyday risk and never more so than during a *cold* war. War no longer belonged exclusively to the domain of armies and soldiers, bureaucratic planners and civilians alike played a crucial role in the imaginary war. Though motivated and legitimised by the unique, extraordinary and exceptional (and imagined) disaster, civil defence was, in fact, a part of (real) everyday existence for many citizens and a practical, daily and ordinary activity for volunteers and professional civil defenders. It was (gendered) civic engagement as well as a professional business, as Hogg (Chapter 4), van Lente (Chapter 6) and Cronqvist and Grant (Chapter 9) make clear. Furthermore, as the Cold War progressed, we see the contours of a new sociotechnical imaginary emerge where nuclear war is replaced by an all-hazards perspective. In this imaginary, a bracketing of the most apparent low-risk/high-impact catastrophe gives way to a more limited conception of emergency preparedness focused on versatility and application across a wide range of scenarios, as the chapters by Molitor (Chapter 3) and Bennesved and Sylvest (Chapter 5) discuss. Though we still live in the nuclear age, this remains the central focus of European preparedness organisations.

Preparedness Redux

Among the questions that lie at the core of this book, one is of special contemporary relevance. What knowledge can we attain from civil defence history to the benefit of our contemporary European crisis-ridden societies as they prepare for future disruptions in the areas of health, military conflict, climate change and digital infrastructure? The association of all things atomic with the Cold War era may delude us to think nuclear war is a bygone threat, but that is not the case. There is little doubt that it should be counted among our present-day risks. Notwithstanding renewed campaigns for nuclear disarmament and the adoption of and entry into force of the Treaty on the Prohibition of Nuclear Weapons, the prospects of averting nuclear war through this route currently look dim. Indeed, some nuclear-weapon states are expanding their arsenals, while others have embarked on huge modernisation efforts that amount to a 'qualitative' nuclear arms race. Against this background, it is hardly surprising that in January of 2021 the Science and Security Board Bulletin of the Atomic Scientists kept the symbolic Doomsday Clock at 100 seconds to midnight—closer than it has ever been. The board noted in its statement that 'an extremely dangerous global failure to address existential threats [...] tightened its grip in the nuclear realm in the past year, increasing the likelihood of catastrophe'.[2]

Still, the main reference point in the resurgence of preparedness as both planning objective and cultural capital is not nuclear apocalypse but a plethora of risks revolving around human interactions with technology and nature. Disruptions likely to follow in the wake of accelerated climate change, cyber warfare or increasingly perilous pandemics have prompted contemporary scholarship and policymaking to focus on the concept of *resilience*. In many respects, these threats differ from that of nuclear war, and as the chapters in this volume amply demonstrate, there was much more to civil defence and preparedness during the Cold War than nuclear war (even if it loomed in the background most of the time). Today, resilience is typically understood as an ability to recover a lost equilibrium (or attain a new one) following a large-scale disruption, shock or catastrophe. As Sulfikar Amir has argued, current debates about resilience are marked by a division between two approaches, one stressing the social, cultural and political dimensions and preconditions of this ability, and the second stressing the need for improving and fortifying its technical and material preconditions. Yet, Amir argues, 'resilience is essentially sociotechnical', which means that is best viewed as 'an embedded feature that comes out of a hybrid construct where individuals and communities are blended with the materiality of technology' (Amir 2018: 4). Arguably, this hybrid understanding informs several research and policy initiatives in Europe. The new EU research and innovation framework Horizon Europe, a €95.5 billion programme running from 2021 to 2027, is one clear example of this trend.

What appears to be still out of sight, though, is that twentieth-century civil defence provides the most complete historical case of such large-scale attempts to make preparedness second nature and to make societies resilient. As the chapters in this book demonstrate, civil defence was also thoroughly sociotechnical. Given their inoculation to a mode of thinking that emphasises the central role of context and circumstance in understanding human life, few historians believe that it is possible to transplant historical lessons from one period to another. Neither do we. Our point here is not that Cold War civil defence provides a template for contemporary systems of preparedness. The insights that history provides are of a different order, but no less important. A core observation emerging from this book is that preparedness measures are not independent of their political, social and cultural context. They affect, limit or promote particular forms of governance, democratic and civil engagement, social norms, identities and visions of how we ought to co-exist as a society. Historical inquiry is as much a process of imagining or discovering the less familiar and expanding our geographical, cultural and temporal reach as it is anchored in the social and political present. If we let historical knowledge of our sociotechnical past inform present-day concerns, including discussions and debates on threats, fears, resilience and safety, we will be in a better position to make sustainable, complex and mature decisions for our desirable futures. We hope this book has contributed to just such an expansion of horizons.

Notes

1. Among the plethora of recent studies that deploy the concept of sociotechnical imaginaries, see, for example, Braun and Randell (2020), Graf and Sonnenberger (2020), Gugganig (2021), Hassan (2020), Scholin (2020), and Sovacool et al. (2020).
2. Science and Security Board, Bulletin of the Atomic Scientists, ed. John Mecklin, 'This is your COVID wake-up call: It is 100 seconds to midnight – 2021 Doomsday Clock Statement', available at https://thebulletin.org/wp-content/uploads/2021/01/2021-doomsday-clock-statement-1.pdf [accessed 29 March 2021].

References

Amir, Sulfakir. 2018. Introduction: Resilience as sociotechnical construct. In *The sociotechnical constitution of resilience: A new perspective on governing risk and disaster*, ed. Sulfakir Amir, 1–16. London: Palgrave Macmillan,

Bourke, Joanna. 2006. *Fear: A cultural history*. London: Virago.

Braun, Robert, and Richard Randell. 2020. Futuramas of the present: the 'driver problem' in the autonomous vehicle sociotechnical imaginary, *Humanities & Social Sciences Communications* 7 (163): 1–10. https://doi.org/10.1057/s41599-020-00655-z.

Graf, Antonia, and Marco Sonnenberger. 2020. Responsibility, rationality and acceptance: How future users of autonomous driving are constructed in stakeholder's sociotechnical imaginaries, *Public Understanding of Science* 29 (1): 61–75.

Gugganig, Mascha. 2021. Hawai'i as a laboratory paradise: Divergent sociotechnical island imaginaries. *Science as culture*. https://doi.org/10.1080/09505431.2021.1884217.

Hassan, Yousif. 2020. The politics of sharing: Sociotechnical imaginaries of digital platforms. *Information Polity* 25: 159–176.

Jasanoff, Sheila. 2015a. Future imperfect: Science, technology and the imaginations of modernity. In *Dreamscapes of modernity: Sociotechnical imaginaries and the fabrication of power*, eds. Sheila Jasanoff and Sang-Huyn Kim, 1–33. Chicago: Chicago University Press.

Jasanoff, Sheila. 2015b. Imagined and invented worlds. In *Dreamscapes of modernity: Sociotechnical imaginaries and the fabrication of power*, eds. Sheila Jasanoff and Sang-Huyn Kim, 321–341. Chicago: Chicago University Press.

Jasanoff, Sheila, and Sang-Huyn Kim, eds. 2015. *Dreamscapes of modernity: Sociotechnical imaginaries and the fabrication of power*. Chicago: Chicago University Press.

McEnaney, Laura. 2000. *Civil defense begins at home: Militarization meets everyday life in the fifties*. Princeton: Princeton University Press.

Schiølin, Kasper. 2020. Revolutionary dreams: Future essentialism and the sociotechnical imaginary of the fourth industrial revolution in Denmark. *Social Studies of Science* 50 (4): 542–566.

Sismondi, Sergio. 2020. Sociotechnical imaginaries: An accidental themed issue. *Social Studies of Science* 50 (4): 505–507.

Sovacool, Benjamin K., et al. 2020. Imagining sustainable energy and mobility transitions: Valence, temporality, and radicalism in 38 visions of a low-carbon future. *Social Studies of Science* 50 (4): 642–679.

Urry, John. 2016. *What is the future?* Cambridge: Polity Press.

Index

M. Cronqvist et al. (eds.), *Cold War Civil Defence in Western Europe*,
https://doi.org/10.1007/978-3-030-84281-9

GPSR Compliance
The European Union's (EU) General Product Safety Regulation (GPSR) is a set of rules that requires consumer products to be safe and our obligations to ensure this.

If you have any concerns about our products, you can contact us on

ProductSafety@springernature.com

In case Publisher is established outside the EU, the EU authorized representative is:

Springer Nature Customer Service Center GmbH
Europaplatz 3
69115 Heidelberg, Germany

www.ingramcontent.com/pod-product-compliance
Ingram Content Group UK Ltd.
Pitfield, Milton Keynes, MK11 3LW, UK
UKHW020908300726
14119UKWH00031B/115

9783030842833